SIDE GLANCES

THE BEST FROM AMERICA'S MOST POPULAR AUTOMOTIVE WRITER

PETER EGAN

ISBN: 9781783182008

Brooklands Books R&T

First published in 2006 by Brooklands Books Limited, an imprint of Brooklands Books, P.O. Box 904, Amersham, Bucks, HP6 9JA, UK.

All Brooklands Books titles are also available on-line at brooklandsbooks.com

ISBN: 9781783182008 (SC) previously 9781855207288 (HC)

On the cover: Photo by Peter Egan
On the back cover: Road & Track columnist Peter Egan, photo of the author.

Inside photography: © Road & Track Magazine and Brooklands Books Ltd.

Peter Egan's childhood in Elroy, Wisconsin, taught him self-control. His academic days at University of Wisconsin taught him self-examination. And his U.S. Army service in Vietnam taught him self-reliance. Once through these life's passages, he tossed them all aside for a career in automotive journalism. Here, "automotive" is meant in the broader sense of two- as well as four-wheel vehicles, with the occasional trike in there for variety. Peter's sense of "journalism" is uniquely—and wonderfully—his own.

Peter and his wife Barb live near Madison, Wisconsin, with a houseful of dogs and cats, and a garage filled with motorcycles, cars and the sound set for his garage blues band.

Peter's Side Glances column has been a first-read for *Road & Track* fans for more than two decades. Whatever the topic— a race car needing exercise, a restoration project gone berserk, a thoughtful chat with his wife—Peter has a way of touching each of our automotive-enthusiast souls.

"I've felt *exactly* that way," readers respond. And so will you.

This collection includes an even four years' worth of Peter's Side Glances columns, from July 2002, "The Un-Original Owner," through June 2006, "A Guide to Collectible Sleepers of the Lowest Rank." Six special features complete the collection, including "A Jaguar in Moose Country," "The Road to Sebring," and Peter's chillingly precise analysis of the Zamboni 500.

Your favorites are here, all to make you chuckle anew and read them to the person nearest in sight. If, through some less-than-propitious reading habits, you've missed any of them, then all the better.

Titles by Peter Egan

"SIDE GLANCES" Volume One

The first book in the "Side Glances" series covering the initial eight years of Peter's association with *Road & Track.*

ISBN 9781855205666

"SIDE GLANCES" Volume Two

A companion book in the series covering the years 1992 to 1997 of Peter's *Road & Track* "Side Glances" columns.

ISBN 9781855205673

"SIDE GLANCES" Volume Three

The third book in the series spanning the years 1998-2002.

ISBN 9781855206367

To my Dad, Phil Egan, who made me build my own gokart, mistakes and all.

CONTENTS

SECTION II
THE FEATURES

Foreword

by Thos L. Bryant
Editor-in-Chief *Road & Track* magazine

The fact that you are holding this book in your hands is testimony to your interest in automotive literature. And when it comes to reading about cars, no one engenders more reader response than Peter Egan. We have been colleagues for many years, and nothing gives me more pleasure than to introduce other car enthusiasts to Peter. For when you read his stories, you are getting to know him. He is precisely the person you imagine as you read about his latest restoration, late nights in the garage, or chasing after another "barn car" that may be something special.

Peter and his wife, Barb, have become friends with thousands of people who have never met them face to face, thanks to Peter's uncanny ability to write from his heart and soul. From his childhood in Elroy, Wisconsin, to his University of Wisconsin days in Madison, to his service in the army in Vietnam, to his life-long love affair with cars, his writing takes you along with him. And it's always worth the trip.

So settle into your favorite chair and prepare yourself for the automotive ride of a lifetime with the incomparable Peter Egan.

Every mile a memory,

Thos L. Bryant
Editor-in-Chief
Road & Track

Section I
THE COLUMNS

July 2002

The Un-Original Owner

After a trip to the doctor's office the other day (nothing serious; I just poked myself in the eye trying to read the small print on my AARP card), I found myself, miraculously, only 27 blocks from a Volkswagen dealer that was not on my way home, so I decided to waft over there with my 1988 Park Avenue and pick up brochures on the Golf turbodiesel.

Funny business, this. I used to hate diesels because they were noisy and slow, and whenever I followed one down the highway and breathed all that soot, I got a splitting headache from the smell. My opinion of these clankers was poor.

But diesels have gradually gotten quieter, cleaner and faster (with turbocharging), and I am naturally intrigued by the 50-mpg concept in these times of Middle Eastern high jinks. Also it doesn't hurt that, in my relentless search for a larger cruising sailboat we can't afford, I've been tutoring myself with a book called *Marine Diesel Engines; Maintenance, Troubleshooting and Repair* by Nigel Calder. He makes the thermal efficiency of these engines sound so compelling, I am temporarily dazzled by logic.

The cover blurb on the book, incidentally, says, "Be your own diesel mechanic."

Frankly, I would much rather Barb be my own diesel mechanic, while I stand on deck and fiddle with a sextant or something, but she seems disinclined to read this book, even though I've pointedly left it on her bedside stand. She continues to read books with titles like *The Cat Who Coughed Up a Furball* and uses the diesel book as a coaster for hot chocolate. This may be a gender thing.

But back to cars. The glamorous allure of diesel engines aside, I also have a soft spot for nicely styled hatchbacks and small sports wagons. Hence the trip to the VW place. I don't need a new car right now, with only 140,000 miles on the Buick, but it's fun to look around. And look I did.

A very polite salesman at the dealership found me a brochure, showed me a few cars, gave me his card and said he'd be glad to help if I made any decisions. While we talked, I sensed that one of his two eyeballs was wandering slightly as it appraised my mildly dented Buick out in the parking lot. Perhaps he wondered if he had a sudden foreign-car convert on his hands.

"The cars were all new. The owners were not wearing coveralls or safety goggles, and there were no Band-Aids on their hands."

I didn't let on that I'd once worked as a service manager at a Volkswagen shop. I've found that you always learn a lot more by feigning total ignorance than you do by attempting to demonstrate partial knowledge. And feigning ignorance is something that seems to come naturally to me. Too naturally, some would say, and without nearly enough feigning.

Anyway, the salesman was helpful and informative and I left with my sales literature in hand. Driving home, however, a vague sense of guilt descended upon me, as usual.

Why guilt?

Well, the truth is I so seldom buy a new car that I hate to use up a salesman's time. It feels dishonest, like a high school kid pretending to be in the market for a new 40-ft. yacht. Or a motorjournalist shopping for a Learjet. ("Better be nice to him, Bob. He might be an eccentric billionaire computer tycoon with a bad haircut....")

On the drive home, I tried to remember how many times in my many decades of car-saturated life I'd gone to a car dealership and actually rewarded a salesman's efforts by driving away with a shiny new car. Shortly, I came up with a total that did not tax my modest math skills:

Once.

When I bought a new Ford Van, in 1995.

Okay, that's not quite the whole picture. My wife Barbara has bought three new cars since we were married, a 1980 Datsun 210, a 1985 Mazda RX-7 and a 1995 Jeep Grand Cherokee. I was present at the delivery of all three, so I got to experience the ritual of the closing, and the heady pleasures of new-car smell, tires with tread on them, etc. And, of course, I got to drive all these cars occasionally. But they were not mine. Barb and I, you see, have entirely different car-purchasing philosophies:

She buys cars that can be driven around, while I tend to buy cars that are disassembled and stored in rotting boxes with the bottoms falling out. I usually bring them home in the back of my van.

Or else I limp home in cars that barely run and then take them apart and store them in my own rotting boxes. Until such time as they can be fully restored and returned to their former glory. Depending, of course, on your definition of "glory."

I reflected upon this unsettling pattern as I drove along and, upon arriving home, sat down and listed on a yellow legal pad all the cars and trucks I'd owned. Using Magic Markers in colors of pink, blue and green, I drew lines through them to denote their condition at the time of purchase.

Pink was for basket cases; cars that arrived home in peach crates or in piles of parts stacked on trailers, as if dropped there by a huge electromagnet at a scrap yard.

Blue was for the barely navigating class; cars that lurch into your driveway in a cloud of steam and rod-knock and are immediately disassembled to see how bad they really are.

Green was for cars that pretended to be running okay at the time I bought them, but eventually needed major work.

It was not a pretty picture:

Of the 39 cars and trucks listed (not counting Barb's three), there were 17 pink ones, 10 blue and seven green. A further four were so good they needed only minor repairs. The Ford Econoline van was the only new one.

So. Thirty-nine 4-wheeled vehicles (we won't get into motorcycles and airplanes here). Only one of them new. Quite a record.

I'm surprised car salesmen don't shoot my tires out when they see me pull into the parking lot. "Get out of here, Egan! We've got things to do!"

Still, I escaped again with my latest new-car brochures. And when I got home, I sat back in my chair, got a cup of coffee and reviewed the features and options of the Golf and the Golf TDI.

It was a nice brochure, full of bright, full-color pictures. The cars shown were new and shiny, with casually stylish young people in the 30-something age group cavorting around them, playing with umbrellas and garden hoses. The architectural backgrounds were very European, with narrow streets, stone buildings and signs in French, perhaps from Paris or Quebec.

There were no wrenches, bead blasters, engine hoists, welders or paint sprayers in any of the pictures. The cars were all new. They didn't need any work. Their owners were not wearing coveralls or safety goggles, and there were no Band-Aids on their hands. People just bought these cars and drove them places. Like Barb.

I put the brochure down and gazed into space for a while.

Then I walked out to my workshop and contemplated my various project cars, their parts scattered here and there on several benches and shelves.

Was it really possible that a person with my love of cars had actually reached the age of 54 without ever buying a new one? An actual *car*, that is, not a tow vehicle or a truck for hauling parts, but a regular, honest-to-God automobile that was unused and all in one piece. Was I to pass through this entire life without ever being known as the Original Owner of a single car?

What was I waiting for?

I sat on my shop stool and gazed at my nearly-restored E-Type for a while and realized that perhaps I was waiting for something new and equally beautiful to come along. Saving myself, as it were, like an aging bachelor with high romantic standards. If I were going to be the owner of just one new car in my life, it seemed only right that it should be something as glorious and perfectly desirable as the E-Type had been in its time.

Maybe I was waiting to be seduced, as I was when I saw my first E-Type. Or my first MG-TC or my first 289 Cobra....

Could it happen again?

Possibly not. Innocence is hard to recover, and I'm not in high school any more.

Maybe all the beauty and romance I'd ever need was contained in the Jaguar, and I could settle for mere wisdom and good fun in my next car. The best of all worlds.

August 2002

Rental Cars of the Mayans

When Barb and I awoke on Easter Sunday morning, we found ourselves in a motel room near the Milwaukee airport. We had flown in late the previous night from a one-week spring vacation to the Yucatán, and had decided to stay overnight near the airport rather than try to drive home in the wee hours of the morning.

Barb got up and went to the motel room window, opening the curtain slightly. "OOOOH NOOOO!" she wailed loudly, like Mr. Bill falling off a cliff.

"What's the matter?" I inquired, opening one eye experimentally.

"There's no blue water and white sand out there! No palm trees or bright sunlight!" she cried.

"What do you see?"

"An idling semi and one wall of a Tyson chicken factory," Barb said. "Also, it's starting to snow."

"Welcome home," I mumbled into my pillow.

As we drove home toward our little place in the country, the snow got worse for a while, then turned to rain. A huge wet American flag at a restaurant along the highway snapped in the chill wind like a gunshot. It was a bleak day.

"Well," Barb said after many miles of mutual gloomy silence, "at least our car is big. You look like you're 10 feet away."

I looked over at her and saw she was right. After the close quarters of the car we'd rented in Cancún, the bench seat in the front of the Buick felt like the veranda of a Southern mansion. We lacked only mint juleps. Such luxury.

If our car in Mexico had been a little tight on space, it was my fault, as usual. When Barb called the travel agent before our trip, she asked me, "What size rental car do you want?"

"The smallest and cheapest thing they've got," I said reflexively.

Later, on the flight to Mexico, I found myself wondering why a person like me, whose whole life is tied up in the romance of cars, was automatically programmed to request a rental car from the infamous "S-Box" category. Why not pay a little extra and order something sporty and fun, such as a Mustang GT, or a big and roomy Lincoln or Cadillac?

After all, I would never walk into a restaurant and tell the waiter, "Give me a small portion of something extremely unappetizing." So why order the cheapest and smallest rental cars?

I pondered this question for a while and decided the answer was simple:

I couldn't bring myself to spend much on a car I didn't own.

Why blow money on a larger rental car (intoned my internal logic) when the difference in price could go toward something really exciting, like a rebuild on the Hewland gearbox in the Lola? Or a nice set of Konis for the Jaguar?

The fact is, I was always ready to spend absurd amounts of money on quixotic restoration and race-car projects, but not a few bucks extra for elbow room and style on a full week of vacation. In the classic economic choice between guns and butter, I always chose guns.

And, in this case, the full extent of our disarmament was revealed right after we landed in Cancún.

Barb and I got off our plane and waited a short time for our friends Rich Mayer and Marlene Piat from Sedona, Arizona, who were flying in to join us on vacation. Rich races a vintage Porsche 914-6, so he, too, suggested we share "the smallest, cheapest rental car they've got." Great minds think alike.

A representative from Budget (nice name) led us by foot a short distance to the rental lot, and, after signing a few papers, we found ourselves presented with a humpy little object known in Mexico as a Chevy Monza. Metallic green, 4-door, automatic, 1.6-liter 4-cylinder engine. It had the sleek lines of a guppy doing a Charles Laughton imitation.

Don't ask me if this car is based on anything we have in the U.S. I'm not one of those motor journalists who can peek at the underside of a car and say, "Oh, yes, this is built on the shortened platform of the same D-body used in the Smackfire minivan, but with a Korean engine they share with Opel in South Africa."

This car said "Chevy" on the trunklid (not Chevrolet, but "Chevy"), and if it had said "Nash," I would have believed it. I don't lose much sleep over cars of this ilk, nor do I dream of them when I do sleep.

In any case, it wasn't a bad little car. Reasonable power and acceleration, a nice tall overdrive for highway cruising, good head and leg room for four adults. Also, the trunk was just big enough to handle all our luggage, with a little careful packing. Both my modest valise and Barb's usual wardrobe trunks borrowed from the Metropolitan Opera fit in nicely.

"We decided the gas pedal had also been painted on, along with the image of a huge Corona truck in our rearview mirror."

But, except for having air conditioning, it was a stripper in the extreme. No hubcaps or switch for the dome light; plastic plugs where the grab handles over the doors would go, etc. Even the radio played only the cheapest forms of Mexican music, without enough trumpets in the horn section.

Also the air conditioner had a mind of its own. It would suddenly shut off and, just when we were about to expire from the heat, cut in again.

"Maybe it has a heat sensor that cuts out when the engine gets too hot," Rich suggested.

I shrugged and pointed to the instrument panel. "The temperature gauge hasn't even moved off cold."

"That's because the needle is painted on," Rich said.

This comment set off an explosion of "painted on" jokes within the car. We decided the gas pedal had also been painted on, along with the image of a huge Corona truck in our rearview mirror.

We spent a night in Cancún and then drove the next day to the

Mayan ruins at Chichén Itzá. The Yucatán peninsula is mostly flat as a pancake, a vast limestone shelf with scrubby, low jungle clinging to the bedrock on a thin layer of soil. How this area provided enough food for the great Mayan cities is beyond me. Some archaeologists say it didn't, and that's why the cities were abandoned. Nevertheless, some 4 million Mayans still live in this part of Mexico, speaking the same language as their forebears who built the pyramids. These days they build good roads; the highways are straight and in good repair.

We toured the ruins all afternoon, climbing the great pyramid and marveling at the energy it would take, in this land of heat and humidity, to carry even one of its great stones to the top, as the Mayans had no draft animals or metal tools and had not yet invented the wheel. Interestingly enough, they had invented a complex calendar, far superior to our own—carved onto a large flat disc. All they needed was four calendars and two axles. And maybe raised white lettering for the months of the year.

That night we stayed at a motel nearby. Under the palm trees was a nice swimming pool with a natural limestone floor full of small caves and hollows, and next to it were grass-roofed shelters with hammocks and an outdoor bar where margaritas could be crafted.

The rest of the week we spent in Mérida, a nice old colonial city somewhat marred by too much traffic and diesel smog, then on to the beautiful and mystical ruins at Uxmal, my favorite stop of the trip. Thursday found us back at a big hotel on the beach at Cancún, which is sort of like Miami but with lower water pressure, where we spent three days swimming and walking on the lovely white sand beach.

The day before we left, Rich confided in me that he'd had a great time in Mexico, but was becoming anxious to get back to his "stuff."

"Me, too," I said.

His stuff, incidentally, is almost exactly the same as my own: guitars, amps, motorcycles, car projects and a workshop full of tools.

On Saturday, we returned our faithful Monza to the Budget lot, said our goodbyes to Rich and Marlene and boarded different planes for the flight north.

When Barb and I got home on Easter Sunday, I found a message on our answer phone from my friend Pat Prince, of Prince Race Car Engineering in Stirling, Illinois.

Pat said he'd found some additional problems with the Lola 204 frame he was repairing for me, and wanted to know if I was willing to spend the extra money to replace a few questionable frame tubes.

"Sure," I said grandly. "We might as well fix it right."

Nothing like a small, cheap rental car to make you feel rich. When you don't blow all your money on *mantequilla*, there are always pesos left over for *las armas*. Rental cars, however great or humble, will never be a part of our stuff.

September 2002

TVR Short Course

"What do you know about TVRs?" the voice on the phone inquired.

It was my old friend and foreign-car repair shop owner Chris Beebe, whose house is just across a small carp-infested creek from our own.

Chris lives so close he could just as easily have shouted a question like this from his back porch, but we've both found that the telephone adds an element of conversational subtlety. Also, it doesn't spook the carp.

"I know very little about TVRs," I confessed. "Small volume British specialist sports car with fiberglass coupe bodywork, sturdy tubular steel frame, independent suspension and a variety of engines. Which, of course, you already know."

Most of that, I told him, I'd learned from looking at a TVR Vixen that our mutual friend in California, John Jaeger, was restoring as a vintage racer. "John is Mr. TVR," I told Chris. "He has a whole library of TVR books and magazine articles."

"Maybe I'll give him a call," Chris said. "A car collector in Beverly Hills has asked me to drive up to Green Bay and do a buyer's inspection on a TVR Griffith. I've got Graham Robson's book on TVRs, but I'd like to do a little more homework on chassis numbers and production dates."

Chris called back that evening and said John had been a great help, and also hooked him up with a TVR specialist in California named Steve Ferron, who was a genuine authority on the cars. "I'm driving to Green Bay to look at the car tomorrow. Want to go with me?"

"Let me consult my busy calendar," I said.

I swiveled my executive-style office chair away from the computer screen and noticed my Aviation Legends wall calendar had fallen down behind my simulated oak veneer credenza.

"Looks like my schedule is clear," I said.

In the morning, Chris picked me up in his blue Miata.

He could also have picked me up with his Healey 100-4, MG-TC, Citroën 2CV truckette, VW microbus or BMW Isetta, but I was glad he chose the Miata. Green Bay is three hours away, after all. And that's by Miata. It's eight or nine hours in a 2CV or an Isetta, and then you have the extra expense of a motel room and postcards.

As we hummed along the four-lane highway at 75 mph or so, I looked around myself at the Miata and said, "This thing is really running nice. Still feels like a new car. How many miles do you have on it?"

"Just turned 140,000," Chris said.

"Any trouble with it?"

"No."

After a while I started chuckling to myself, and Chris asked what was so funny.

"I was thinking about my first TR-3..." I told him.

Then Chris started to grin, and pretty soon we were both chuckling. The concept of 140,000 miles without trouble is always a source of mirth to British car guys.

When I calmed down a bit, I began looking at the TVR book Chris had brought along. There I was reminded that TVR got its name from Trevor Wilkinson, who used selected letters from his first name for the small company he founded in Blackpool just after World War II. Later, he built a factory in a place called Hoo Hill. Only in England.

> ***"The concept of 140,000 miles without trouble is always a source of mirth to British car guys."***

Like so many of these limited-production British sports cars, the TVR was part of what might be called the Fiberglass Revolution, along with Turner, Elva, Lotus, Ginetta and others. The designer could produce an inexpensive body full of complex curves without the need for skilled panel-beaters to make every car, then bolt it to his own chassis, using proprietary engines and suspension parts from Ford, Austin, MG, etc. Some of these cars were quite beautiful. Others looked like catfish.

TVR built a long series of cars—Grantura, Tuscan, Vixen, and so on—mostly short-wheelbase coupes with a strong familial similarity. The Griffith came about when an American TVR buff named Jack Griffith decided to "do a Cobra" on the TVR Mk III and shoehorn a Ford 289 into the engine bay, replacing the MGB engine. He worked out a deal to have complete cars, sans engine and gearbox, sent to the U.S. These would then have Ford drivetrains installed and be marketed only as "Griffiths," leaving the TVR out of the name. A young man named Mark Donohue did some of the early test driving.

Early Griffith 200s used a Ford 289 rated at 195 bhp, then a hotter Griffith 400 was introduced (400? 400 what? Why not 800, while we're at it?) with the 271-bhp K-code engine from the Cobra and the body revised with a Kamm-style rear end. According to Robson, about 300 Griffiths were made between 1963 and 1965. He says they quickly earned a reputation for indifferent construction, but were shatteringly fast, with a top speed of about 155 mph. The car Chris and I were going to look at was a Griffith 400.

We found the owner's home in a tidy new suburban section of

Green Bay. When we arrived the Griffith was sharing an immaculate four-car garage with only a lawnmower and a gas can. No tools, clutter, grease, oil slicks, spare engines or old parts cars to be seen; just a very clean Griffith painted a Cobra metallic blue, with red and white racing stripes. Nice-looking car. Neat red interior.

The car's owner, a pleasant fellow named Tom, produced a floor jack so Chris could examine the underpinnings. He tapped and probed for about an hour and found the car to be in pretty good shape underneath, except for a couple of thin and rusted bottom tubes in the chassis.

With the car back on the ground, Tom and Chris took a drive. The Griffith started up with a throaty roar, catapulted out of the garage and was gone. I sat in the Miata and amused myself reading about TVRs. When the car came back, Tom asked if I would like to take a ride, so I climbed into the passenger side (long, roomy footwell, good bucket seats, narrow shoulder space) and Chris motored out onto the highway. The cockpit instantly filled itself with wonderful Hi-Po Ford noises. And haze.

"How is this thing?" I asked loudly.

"Great engine, kind of a stiff clutch, pretty good suspension and steering. No ventilation, though. Lots of smells and heat from the engine bay."

That pretty much summed it up. When Chris put his foot in it, the car accelerated as only high-performance cars from the 1960s accelerate—a smooth yet edgy mixture of steel drum music, rumble and thrust, with back-flattening torque from idle to redline. Great engine indeed.

More surprising was the ride and suspension compliance, which

was quite civilized. It reminded me of my friend Tom Cotter's 289 Cobra, firm and controlled, but not punishing. Chris took it through a fast sweeper, and it seemed planted and flat. The short wheelbase, however, had me wondering about the feasibility of more speed. It seemed like a car that could stand about 6 in. more wheelbase, shades of the Sunbeam Tiger. Nevertheless, the Griffith felt compact, low and agile.

Chris was right about the ventilation. No air movement, plenty of greenhouse area to soak up the sun's heat and a heady mixture of engine compartment smells that were curiously half British and half American—an aroma of high-octane fuel mixed with a whiff of V-8 oil breather outlet, old British wiring loom and wool carpet.

"If I owned this," Chris said, "I'd have to install side vents in the fenders and some kind of exit vents at the rear. Also seal off the firewall a little better."

We pulled into the driveway, Chris wrote some more notes for the prospective buyer in California, then we said our goodbyes to Tom and climbed in the Miata and headed for home.

The Miata felt a bit down on power after the Griffith, but the cool air coming through the vent system felt good. I found myself breathing deeply again. And glad to be living in a time when we can enjoy sports cars from more than one era. Moving from the Griffith to the Miata was like flying your last mission in a Thunderbolt and going home in a Beech Bonanza.

When Chris dropped me off at the end of my driveway that evening, I opened the mailbox and found a new *Hemmings* inside.

I took it to the house, microwaved a cup of the morning's leftover coffee and sat down on our porch swing to read. I turned immediately to the "Ford Mustang Cars For Sale" section;

For my tastes, the Griffith was a little too demanding as a daily driver, but riding in it reminded me once again that I am destined to own something with a 289 Ford High Performance V-8 in it.

The Gypsy woman said so when I was born, and I believe she specified a 1965 or 1966 Ivy Green or Dark Green Metallic K-Code GT Fastback with a 4-speed transmission.

October 2002

Garage Cleaning Tips

Just last week I got a call from a man named Craig Constantine at the History Channel who asked if they could interview me for a program on the history of the motorcycle. At first I was flattered, of course, but then the downside hit me: I've been alive for about half the history of the motorcycle!

Soon I started to wonder if they really needed an interview or merely wanted me cataloged as a venerable artifact. "Here are some Roman coins dredged up from the Aegean Sea, and here's Peter Egan wearing the Bell helmet he bought in high school." That sort of thing.

Ah, well, it would be fun either way. I like history, and the History Channel too, especially any program that shows fighter planes from World War II or Hitler looking worried while examining a map on a huge table, so I agreed to the interview.

"We'll have a film crew in Milwaukee next week, doing some things with Harley-Davidson," he said. "Do you live near there?"

"About an hour and a half west," I said, "out in the sticks."

"Good. Maybe we could go to your place and do the interview in your workshop."

"Fine," I replied, with my voice trailing off.

My workshop? Suddenly the hair stood up on my head like a thinning gray fright wig. My garage was a mess. One look around and they'd pencil me in for a History of Swine. Something had to be done.

So last Saturday, on a beautiful warm summer morning, I went out to the garage with a pot of strong coffee and three dogs, threw both doors open and began to clean up. I worked relentlessly all day and into the evening. By midnight, the place was spotless.

In fact, the cleanup was so successful, I thought it might be useful to offer a few helpful Martha Stewart-like tips to workshop owners who are thinking of doing the same thing. If you're like me, you tend to let things slide until they are almost intolerable, then attack the problem with a vengeance. I've done this so many times now, I have a regular regimen that allows me to whip the garage back in shape in only a single 17- or 18-hour day.

Here's the program:

1. First, clean up your band corner. Our particular garage band practices every Sunday afternoon on a luxurious 12 x 12-ft. piece of indoor/ outdoor carpet, where we have drums, amps, mic stands and a full Peavey sound system set up. This is theoretically a simple area to clean up, unless your band is like ours and drinks beer. A lot of beer.

"Don't ever launder coveralls with any of your wife's clothing unless you have a death wish or need anecdotal material for cocktail parties."

On a hot summer night, the six guys in our band consume about as much beer as an American Legion Post. Our drummer especially can put it away. Yet he always keeps perfect time, as far as we can tell. But then, what do we know? He probably thinks we keep perfect time too.

In any case, we leave behind a lot of beer bottles. So I pick them up and always dump the dregs outside under the lilac bush (which is flourishing) and fill two or three large shopping bags with empties. You can later take these bags to the dump and put them in the recycling bin while the dogs fight over a pork chop bone or roll in something dead.

Then I vacuum the band carpet and use fresh-smelling lemon Pledge on my Fender Super Reverb amp. This is also a good time to dust the framed portraits of Muddy Waters and Sonny Boy Williamson that hang on the wall nearby. As a final touch, I like to take my shop-vac and clean all the cobwebs off the Texas flag and the Rolling Stones poster. You have to be careful here, or the shop-vac will suck the flag off the wall, and you'll be looking for thumbtacks in the vacuum cleaner basket. A messy job, but a great way to find those old 8-mm sockets and the ball bearings you left out of something critical. Band area done.

2. Next, clean up your motorcycle area. The Ducatis and BMW can usually be shined up with a little spray polish, but a 1960s' Triumph or a Shovelhead-era Harley will usually leave enough oil on the floor to keep us on good terms with our dear friends, the Saudis, almost forever. I used to scatter sawdust on these oil slicks, but now I use electrical contact cleaner and paper towels. These will burn brilliantly when you torch the pile of empty beer cartons in your incinerator.

   Once the oil's gone, you can move all the bikes outside and sweep up the dead flies. Flies that are stuck in thick green gear lube should be whisked away with solvent and paper towel rather than vacuumed up, or the gear lube on the vacuum cleaner attachment will leave streaks on your Rolling Stones poster and Texas flag when you go after the cobwebs.

   British cars, of course, require exactly the same technique, but be careful not to run over the oil slick or you'll leave tread prints all over the floor and your friends will tell British car jokes you've heard before.

3. Reappraise your automotive "spares." Take a good long look through all the boxes of worn-out, grease-caked parts you've replaced on your current car-restoration projects (in my case a Jaguar and a Lola). Be honest with yourself here. Are you really ever going to reuse those cracked old rubber brake lines, rusty ball joints and scored rod bearings again on your freshly rebuilt classic? Of course not. Throw them out!

   On the other hand, you can never be sure, so it's always best to pack them carefully away in a box and carry them down to the dilapidated horse barn for storage, over in the corner with the old Fiero seats that came out of the TR-4 parts car you sold five years ago.

4. Dust the canoe. If you're like me, you've got a 74-lb. wood-and-canvas Atkinson Traveler canoe suspended like the Sword of Damocles from the ceiling in your garage by an insane system of

pulleys and ropes, so if it ever falls down it will probably destroy itself and land on the Jaguar, or kill your drummer. Who may or may not notice. In any case, nothing says "bad housekeeper" louder than a dusty canoe. Get up there on a rickety stepladder with some lemon Pledge and make it shine! If it's summer, of course, you can take the canoe down and use it on those weekends when there's no racing.

While we're on this nautical theme, this would also be a good time to do something with the laminated ash sailboat tiller and steel rudder that are sitting on top of the worn-out Formula Ford tires in the corner. A rudder is a flat thing, and may be stowed nicely in the back of your wife's clothes closet.

5. Put all your fishing equipment, including bait buckets and bluegill stringers, back in the hot trunk of the Buick, where it belongs. A limited amount of fly-fishing equipment may be hung neatly on the workshop wall, as it implies you have class, or at least think you do, even if you can't actually fly-cast without hooking your own hat.
6. Store the spare seats that came out of your 9-passenger Ford Club Wagon in your neighbor Jim's huge, dry, modern, mouse-free basement, promising that you or your heirs will retrieve them at some undisclosed time.
7. Donate the stack of widened chrome 5-bolt Volkswagen wheels to your other neighbor, Chris, who still owns a VW Microbus. Admit that you, personally, will never own another air-cooled Volkswagen product because you've been there and done that. Three times. Incidentally, you won't be able to clean the rust circle off the concrete floor where the wheels sat for 10 years, so you'll want to stack the Formula Ford tires over the stain.

8. Take all old car and motorcycle batteries to the dump for recycling. Put them inside a cardboard box, keep the dogs away from them, and don't rub your eyes. Try to keep the acid burns in your clothes asymmetrical, for a more natural look.
9. Wipe out your parts cleaner and throw all the old oil filters away. Unlike used rod bearings or worn balljoints, they aren't worth saving for future generations.

10. Temporarily store your bicycle shoes and sweaty gloves in the trunk of the Buick with the fishing gear. Dust the bicycle.
11. Take your coveralls to the house and throw them in the washer. Alone. Don't ever launder coveralls with any of your wife's clothing unless you have a death wish or need anecdotal material for cocktail parties.
12. Open a beer, sit back and enjoy the view of a job well done. If you've done a thorough job, it should be about midnight.

That's about it. A few simple clean-up tricks like this can make you the envy of your racing friends and your garage band, and can also make you and your environment appear almost normal during a television interview.

November 2002

The Power of Style

I must admit that our Ford van, fully loaded for a camping and sailing vacation, looked like something Jed Clampett and his family might have taken on the road, if they'd had time to get rid of the old pickup.

Inside the back of this metallic blue cavern we had two bicycles; a Eureka Timberline tent; a boat rudder; a vintage Evinrude 2-hp outboard engine; a small charcoal grill; a toolbox; a grease gun for the "Bearing Buddies" on the trailer; my favorite Swedish camp ax; three coolers filled with food, beer, soda and margarita mixin's; anti-mosquito citronella candles; sleeping bags; a preinflated double air mattress; extendable hot-dog roasting sticks; and enough paper shopping bags of sheer *stuff* to dazzle the logistics officer of the Third Army.

My wife Barbara and I differ slightly in our camping philosophies; I am of the "just give me a knife and a piece of flint" school of woodcraft, while Barb worries that we won't have a proper tablecloth to match each subtle shift in ambient light and foliage coloration. I read Thoreau and she reads *Country Living*. I'm lucky our tent doesn't have drapes. In any event, I get to grumble a lot and pose as an ascetic, while still enjoying the posh camp life of a Bedouin king.

Added to this internal load, our van was also pulling our 15-ft. West Wight Potter, a stout little sailboat with a cabin that actually sleeps two adults in moderate comfort. We bought the boat a year ago because, despite its small size, it has a reputation for handling heavy weather well, and it's what's known in the brokerage business as a "salty" boat. It *looks* properly nautical.

So, essentially, we bought the boat because we liked its lines. Couldn't help ourselves.

Dragging all this impedimenta, we left home and headed for Door County, the "thumb" of Wisconsin that sticks out into Lake Michigan. Door County is the Cape Cod of the Midwest, a peninsula of rocky harbors and lovely old fishing villages that have been restored to the exacting standards of modern tourism. We would be sharing a campsite with our friends Jim and Patty Wargula, whose modest motorhome the size of Willie Nelson's tour bus was already ensconced in the hills above the port of Ephraim.

"Grace of line is not everything, but in the world of cars, boats, motorcycles and airplanes, it's usually the thing that ultimately makes us surrender our money and our hearts."

After four hours of travel, our E-150 Econoline at last labored up the winding road out of town to the campground, and I found myself wondering for the nth time if I should have bought a van with the larger 5.4-liter V-8 rather than the base 4.6-liter V-8. The 4.6 is very smooth and adequately peppy for most driving and towing, but every so often I find myself wishing for just a bit more torque and wallop.

I could have bought a Chevy Express van, of course, with the classic 350 (5.7-liter) V-8 as well, or a Dodge Ram van with a 5.2-liter or 5.9-liter V-8 as well, but we somehow settled on another Ford, despite some nagging problems with the E-350 we traded in.

Why another Ford, with so many other good choices out there?

Because we liked its lines.

We found our friends the Wargulas, enjoyed an evening around the campfire, and in the morning Barb and I towed our little boat down to the harbor at Ephraim and sailed out into the open waters of Green Bay. In a gentle breeze, we made it to Chambers Island (about halfway to Michigan) and back that morning.

In the afternoon, Barb and Patty went shopping for small, costly doodads in Door County's many gift shops, while Jim and I, as usual,

prowled the harbors and marinas in search of interesting sailboats. It is part of my life plan to eventually trade up to something substantially larger than the West Wight Potter for extended cruising on the Great Lakes. And it is part of Jim's life plan to come aboard with Patty and drink margaritas, so he is always helping me look for the right boat.

We stopped in Sister Bay at a place called the Yacht Works and I talked to a salesman friend named Peter Jacobs to see if he had any neat bargain boats lurking around.

"Is that 30-ft. Cape Dory out front for sale?" I asked him.

"No, that's a customer's boat."

"What about the Pacific Seacraft?"

"That's a customer's boat, too," Peter said.

Then he turned and looked at me for a moment and grinned. "You really have a thing for those good-looking 'character' boats, don't you?"

I thought about that for a moment and smiled, somewhat bemused. "Well," I said, "I must admit that I have never purposely bought anything in my whole life that I thought was ugly."

"Good point," he said, laughing. "I guess I haven't either."

We had a great week of sailing and campfires, and then the Wargulas headed home, leaving us alone at the campsite. On Monday,

Barb and I drove back up to Sister Bay and rang the doorbell of Tom and Anne Harrer. Tom is a former SCCA racer (TR-4, Sports 2000) and professional race-shop owner who now resides in Door County.

Tom had just this spring taken delivery on one of the world's most charming and traditional sailboats, a 28-ft. Bristol Channel Cutter—which was built about four blocks from the *Road & Track* office in Newport Beach, California, so of course we had to have a look at it.

Tom not only gave us a look at it, but also took us sailing. A great day of slicing through whitecaps in a 22-knot breeze. The boat, with its long bowsprit, extensive brass and mahogany brightwork was stunning to look at as we sailed along. Sailing this thing was like trying to keep your eyes on the road while gazing upon the instrument panel, headlamps and bonnet of an MG-TC or a Morgan Plus 4. As a place to be—as well as a thing to use—the boat was just about perfect. Which I suppose is true of all boats, cars, motorcycles or airplanes worth having.

We were not the only people who found it beautiful. When Barb and I strolled around the marina at sunset that evening, we noticed that everyone stopped to look at Tom's boat, often for a long time, just gazing and letting its shape and detail wash over them. Like us, they couldn't help themselves.

It has beautiful lines.

Which, of course, is why this traditional old design is still in production.

On Wednesday, our vacation was over, so we broke camp, pulled our boat out of the water and headed for home.

South of DePere, we were driving along when a rather nondescript red car abruptly changed lanes in front of us and I had to brake slightly. Barb looked up from her magazine, watched the car for further signs of unpredictable behavior and then asked, "What kind of car is that, anyway?"

I was mute for a moment as my eyes darted frantically over the back end of the car, trying to come up with some name. "I think it's some kind of Oldsmobile," I said at last, "but I can't remember what that model is called."

"It's so nondescript," Barb said. "It looks like a million other cars on the road."

"It's a far cry from the 1972 Olds Cutlass Jim and Patty had when they were first married," I admitted. "Jim liked the design so much, he went right out and bought a new one. Couldn't help himself. That car had beautiful lines."

On the rest of the return trip I found myself wondering why so many of the world's car manufacturers were content to produce cars whose appearance almost no one admired. Does a beautiful line, drawn on paper, cost so much more than an ugly one?

If you spend millions on engine and transmission development, how many more dollars does it cost to fashion sheet metal that pleases the eye?

Why does the public instantly recognize a handsome automobile, while hundreds of professionals in a design department can't seem to tell bland from beautiful? And wouldn't the people who buy those bland cars be just as content to buy one that was beautiful?

My dad, who had almost no interest in cars, absolutely had to have a first-generation Mustang because he loved the way it looked. An entire generation was dumbstruck by the fragile beauty of the MG-TC and the sleek lines of the Jaguar XK-120, and later the E-Type. The whole world fell in love with the original Mini, and now people are standing in line for the new Mini, while a dealership across the street blows nondescript cars out the door with zero-percent financing.

We can talk all we want about engine design, torque, horsepower, gear ratios, shock absorber damping and the placement of anti-roll bars, but if a car doesn't look right—if its outward appearance doesn't excite the imagination on some level—most of us just don't care.

Grace of line is not everything, but in the world of cars, boats, motorcycles and airplanes, it's usually the thing that ultimately makes us surrender our money and our hearts—and sometimes our good judgment—without a moment of hesitation. And if it turns out later not to have been a very good machine, even our remorse is an endless source of pleasure.

Good or bad, when the styling is right, we can't help ourselves.

December 2002

To Start an Engine

An unwritten rule of auto mechanics states that when you pull an engine, there is always one thing you've forgotten to disconnect. It's usually a choke cable or a ground strap—something just strong enough to swing the engine like a wrecking ball into the firewall, or the nearest painted surface.

Another rule, of course, is that, when you put the engine back in the car (years later), there is always one thing you forget to tighten adequately. Usually it's a water hose or an oil line that spews fluids all over the floor and under your jackstands, or a stray wire that throws sparks like a wall switch in a Frankenstein movie.

Knowing this, I sat on a shop stool in my garage for a very long time the other night drinking a Mountain Dew (plenty of caffeine for staying awake in the wee hours), staring at the engine compartment of my E-Type Jaguar.

The engine was in; everything was hooked up. Time to crank her over. But certainly I'd forgotten something.

Starting with the radiator, I allowed my eyes to travel slowly backward along that lovely, long straight-6, trying to imagine what would go wrong. Throttle linkage connected, ground strap on, fuel lines tight. Still no water in the cooling system, however, and no exhaust system. Just the twin cast-iron headers to direct the hot flaming gases downward and prevent eyebrow fires (always a serious concern in the Egan family).

I have a superstitious belief that adding coolant and hooking up the exhaust system on an engine that has not yet proven itself to have oil pressure is a form of hubris that invites disaster, like buying trophy polish before the Indy 500.

I checked the oil dipstick again. Full, all 10 quarts.

And, yes, the gearbox had oil in it. Only the day before, I'd finally removed the small note over the ignition key that said, in my own handwriting, "THERE'S NO OIL IN THE GEARBOX, STUPID." A procrastinator's trick from the night I installed the engine and transmission, but was too tired to do another thing and drank a celebratory margarita instead of messing around with gear oil and funnels.

I got tired of looking at this warning, so I filled the transmission with golden gear oil at last and threw the note away.

Then I lay on my creeper under the car and waited for gear oil to ooze out of some orifice or seal. But none did. Good. It was spookily dry and clean on the underside of the engine and gearbox, especially for a British car. But probably not for long.

Show time.

I climbed out from under the car and placed the ground strap on the battery and stood back, half cringing, like a Neanderthal man who'd just shaken a spear at a woolly mammoth and was waiting to see what would happen next.

Nothing. No sparks, no fires, no small mushroom clouds from the dash or starter motor. I reached into the car and turned the ignition key to ON. The fuel pump, which had been dormant for nearly three years, instantly began pumping clean new fuel into the glass filter bowl and thence into the three big SU carburetors. The float bowls filled and the fuel pump stopped ticking.

No leaks. No stuck and overflowing needles and seats. Still dry under the car. Unbelievable.

I took the sparkplugs out, disconnected the coil and spun the engine with the starter motor. The oil pressure needle gradually began to tremble and awaken, and then it climbed to 60 psi. Still no oil leaks.

I put the plugs back in, set the choke on cold, sat in the car, checked the gearbox for neutral, turned the key, took a deep breath and hit the starter button.

"I'd worked on the car for so long as a stationary icon of mixed worship and dread, I'd nearly forgotten about the promise of mobility."

The engine turned over smoothly a few times then immediately started running with a deafening roar, like half a Merlin with straight pipes. Fast. Too fast. The revs soared past 3500 and I shut it off.

Of course. I'd forgotten to install the return spring on the front carburetor, because I couldn't find it in any of the dozens of boxes and trays and drawers full of Jaguar parts that had filled my garage these many seasons.

I walked to my treasure trove drawer of random forgotten springs and selected something that probably came off a Briggs & Stratton lawnmower in 1963. I bent the spring into temporary usefulness and hooked it on the throttle shaft.

With the spring in place, I started the engine again and it settled down to a smooth, if loud, idle and it revved with a glassy, almost turbinelike silkiness, the engine hardly rocking on its mounts.

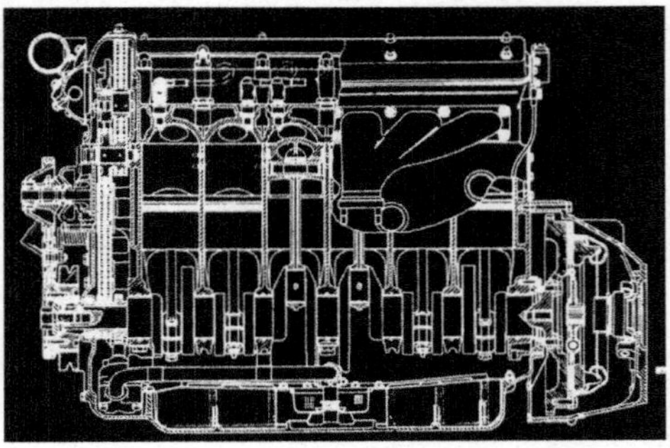

Ah, the joys of a big straight-6. My favorite engines. Nothing like 'em. Lovely balance, wonderful sound. And the Jaguar six is an engine that actually says *varoom!* when you rev it up, as if pronouncing the word in precise English.

I shut the engine off before it had time to overheat, then opened the garage doors into the warm summer night to let the smoke out. The garage was thick with a white cloud that mingled the smells of exhaust, hot paint and gasket cement. And maybe a little human sweat. As personal valet for this magnificent engine in its moment of reawakening, I was soaked in perspiration.

I sat on my shop stool and looked out into the night, where lightning bugs twinkled through the dissipating haze like rising champagne bubbles. A beautiful night for an engine start-up. Especially an engine that ran, idled normally, had excellent oil pressure and leaked no fluids. My brain almost buzzed with pleasure, or perhaps from near asphyxiation from fumes. In any case, I was quite happy.

It would be nice to say that this successful Jaguar engine rebuild was my own doing, but I had, in fact, farmed most of it out to my friend and professional race-car mechanic, Steve Straavaldsen.

I hesitated to do this, because rebuilding engines is actually my favorite part of auto mechanics. But there was a moment last fall when I looked at the immense number of Jaguar parts piled and scattered around my garage and realized if I added an engine rebuild to the endless list of tasks to be performed, the car would remain in my workshop for yet another summer without having its tires touch the ground. So I called Steve.

In the world of automotive work, Steve is a rare paragon of efficiency. I dropped off the engine on a Monday morning, and he called me back that afternoon to say he'd completely stripped and examined it. The crank was fine, but it needed pistons and a rebore. He sent the cylinder head to California for refurbishing at a place called Straight Six, then rebuilt the short block and had the engine back to me by the end of the month. Amazing. All I had to do was install the carbs, crankshaft pulley and distributor and set the timing.

And now the engine was in the car, running. And running beautifully. For the first time this millennium.

Done working for the night, I cleaned my hands, took off my coveralls and sat down to plan my next move. Not much left. The exhaust system, coolant, some heat shielding. Put the seats, carpet and console in. If this were a race car—and I were 21 again—I'd pull an all-nighter and have it done by morning. But it's not, and I'm not, so it would probably be done in a week.

Sitting there and pondering the Jaguar, I was suddenly struck by a whole new vision of the car. With the starting of the engine it had transformed itself from an endless list of tasks to be crossed off into a working car, a machine that might soon provide transportation for two people and their luggage. A magic conveyance in which to go places.

If this sounds stupid and obvious, I apologize, but I'd worked on the car for so long as a stationary icon of mixed worship and dread, I'd nearly forgotten about the promise of mobility, which I'm told is the reason cars were invented in the first place. It took the actual ignition of gasoline by sparkplugs in those six big cylinders to remind me why I'd bought the car, and what it was for. All that smoke and the crackling exhaust were like a bolt of lightning hitting the primordial soup and jolting mere protein and minerals into Life. Soon reptiles would walk the earth. And Jaguars.

At 1 o'clock in the morning, I finally turned out the garage lights and walked up to the house.

I looked in on Barb, who was sound asleep, then went downstairs to the kitchen, where I sliced up a lime, salted the rim of a glass and fixed myself a large margarita on the rocks. I carried the drink into the living room, sat in my favorite reading chair, my clothes still reeking of exhaust smoke, and amused myself for an hour or two by looking at the Rand McNally Road Atlas, one page at a time.

January 2003

Tales of the Dreaded Comeback

A couple of years ago, I took a cross-country trip from Walnut Creek, California, to Charlotte, North Carolina, with my friend and fellow R&T contributor, Tom Cotter, who had just bought himself a red 1965, 289 Cobra. One of my favorite moments of the trip occurred as we were leaving Walnut Creek, buckled in and ready to hit the road.

The mechanic who had maintained the car for its owner walked over, blessed us with a wave of his hand and said, "If you have car trouble, may you have it far from here."

We all laughed (some louder than others), but I knew exactly how he felt.

No mechanic likes to see his work come back to haunt him. And all mechanics privately hope that time, mileage and natural wear will absolve them of responsibility for any future mechanical failures, which are as certain as death and taxes.

Essentially, everyone hates a comeback.

During the years I worked as a car mechanic in the 1970s, one of the worst things in the world was to look out the window and see a car you had recently worked on returning to the shop on a tow-truck hook or a flatbed.

A loud cry of "Comeback!" would reverberate through the shop and your heart would sink. You hoped fervently that the car's latest problem was unrelated to your own work, but feared the worst. Your immediate thought was, What could I have possibly done wrong on that car?

Unfortunately, there was a pretty good chance you would spend the rest of the day finding out, even if the problem was not related to your own work. Sometimes, however, it was.

Sad to report, but the very first car I worked on as a "professional" mechanic returned as a comeback. It was an oil change on a Datsun 1600 Roadster owned by a young woman. The parts department looked up the part number on a chart and handed me an oil filter. I changed the oil, tightened the filter, started the car, checked for leaks (none) and the car left.

"A loud cry of *Comeback!* would reverberate through the shop and your heart would sink."

An hour later, the Datsun came back on a tow hook. The young woman had driven about 20 miles before the oil filter fell off and the engine developed loud knocking noises. She'd noticed a red warning light, but didn't know what to make of it and kept driving until the engine stopped.

As it turned out, I'd been given the wrong filter. It looked like the old one, but the circumference of the rubber gasket was slightly larger, so it wasn't fully seated against the block. The oil filter felt fine when I put it on, but eventually squeezed out its seal.

The parts department took partial responsibility for this little disaster, but my boss and mentor, Chris Beebe, told me I should have matched up the sealing rings to make sure they were identical.

And I did, forever after.

In any case, the shop had to rebuild the engine—this job was given to a more experienced mechanic—and I had to see the Datsun sitting in the shop, forlorn and engineless for at least a month, as a daily reminder of my oversight.

I didn't have too many comebacks after that. The only one of any consequence didn't actually come back. I did a tuneup on an old Volkswagen Beetle and the owner made it to Chicago before his distributor cap fell off—in the pouring rain on the Dan Ryan Expressway on a Friday night. This made him very happy.

He got towed to a shop, where they put the cap back on, but he called on Monday to let us know what he thought of us. Or of me, rather.

As he should have.

I felt terrible. I probably didn't get one of the spring clips on all the way. Should have. Didn't. Maybe the clip was bad. Who knows? Sometimes we make mistakes. A mechanic fastens hundreds of small parts together during a day of work and has to think about every one of them, or things go wrong.

Fortunately, most comebacks at the shop were much less dramatic than this. They were more along the lines of, "This gentleman says you installed the 27-piece exhaust system on his Volvo last month, and now there's a rattle somewhere."

The best ones, of course, were those in which you knew instantly you were guilt-free: "Yeah, buddy. Listen, you replaced my water pump last winter and now the trunk leaks when it rains."

"That's nothing (you are tempted to reply); I had an appendectomy, and my uncle's barn burned down."

Another variety of comeback—especially in the '70s—was the Unsolvable Running Problem. The car industry at large was grappling with new emissions laws, and a lot of cars just ran poorly right out of the showroom. The owner would say, "You did a tuneup on my Austin Marina GT, and it still doesn't run any better."

The cruel answer was, "And it never will. Unless we replace the intake system with the twin SU carburetors and manifold from an old MGB, at great expense. You might want to think about trading it in while the mileage is low."

It was depressing to work on cars you knew would never, ever run right. Especially if you had to work on them more than once.

When I finally quit working on other people's cars in 1980 and became a full-time journalist, I thought my comeback days were over. But they were not.

As a car junkie/hobbyist who has restored at least 10 cars since then, I have discovered a new and insidious form of the comeback. It probably comes under the heading of entropy rather than error, yet still generates the same dread.

It hit me fully for the first time last fall, when I put my 1963 Porsche 356B up on the jackstands to do an overall lube job and oil change before I put the car away for the winter.

While the oil was draining out of the sump, I lay on my back and looked at the underside of the engine, gearbox and axles with my shop light. After three years on the road and two full trips across the U.S. and back, the underside of the car was still almost as neat and clean as it was when I'd finished my restoration. No leaks, no telltale trickle of oil from the sometimes-fragile oil cooler buried deep within the engine's shrouding. The painted parts were still holding paint, and the plated bolts and screws were still plated. The car ran perfectly. And yet the miles were piling up. Sooner or later, something would go wrong.

As I lay on the creeper and looked up at the engine, I found myself saying aloud, "If this engine and gearbox ever have to come out of here again, I don't want to be the guy who does it. I don't want to be around when this car gets old and tired again."

The next day I put an ad in the paper, and shortly thereafter sold the 356. I decided to put all my old sports-car restoration eggs in one basket, so to speak, and use the money to finish my 1967 Jaguar E-Type project.

I told myself that I didn't need two old sports cars to maintain, but the real reason I sold the 356 was the sense that I was gradually driving it toward a comeback, a time when I would have to do the same work over again. And I didn't want to.

This is not an unusual phenomenon. My buddy Pat Donnelly recently sold his beautiful Lotus 31 vintage Formula Ford and bought a Titan Mk. 6C. He told me he did it partly because the Titan is newer and faster, but mostly because he was simply tired of working on the Lotus, dealing with the same bolts, motor mounts, hoses, lines and body fittings over and over again. If he had to pull an engine out again, he wanted to do it on a *different* car. So he sold the Lotus when everything was sorted and perfect. No comebacks, no revisiting of his own work.

My friend Richie Mayer in Arizona recently sold his beautiful 914-6 vintage racing car, and the reasons he gave me were—almost verbatim—the same as Pat's: excessive familiarity and a reluctance to repeat his labors. The old "been there, done that—one too many times" syndrome.

Now, of course, Richie is looking for another old car to restore. Maybe an Alfa. Pat is grappling with starter problems on his new/used Titan, and I am struggling to get the alignment and ride height correct with the torsion bar system on the Jaguar.

We all have plenty of undiscovered car trouble ahead, no doubt, but at least the problems we will eventually conquer are new ones.

For now.

February 2003

Mercury Blues

Here's a weird coincidence, for those who are into the unlikely, if not exactly the occult or spooky:

I was sitting around about two weeks ago, playing my guitar and practicing an old blues song called "Mercury Blues." It's a song originally written by the late K.C. Douglas and later famously recorded by Steve Miller. It's a great song, and our garage blues band, The Defenders, does a version of it that some people say defies description, or even accurate identification. But we do it nonetheless.

It is, basically, a song about a guy who's going to buy him a Mercury because he stole a girl from his best friend, but the fool got lucky and stole her back again because she knowed he had a Mercury.

This kind of thing happened to me all the time in high school because I had no Mercury—among other things—so I play it with a certain amount of conviction, if not skill.

Anyway, there I was, sitting in front of my Fender amp in the garage and hammering away on my Gibson ES-335, rehearsing for our traditional Sunday evening band practice. Taking a break, I set the guitar down and took a sip of Diet Mountain Dew (no bourbon until later in the evening, 'round midnight, when I get that hellhound on my trail) and got thinking about the words to the song.

Had I, personally, ever really wanted a Mercury?

Yes.

I'd wanted a 1949 Merc ever since seeing *Rebel Without a Cause* when I was a kid.

It wasn't just the movie, of course, that made me want one. The 1949–1951 Mercurys were already considered cool cars, James Dean notwithstanding, because they looked so good and came with Mercury's hotter, upgraded version of the Ford flathead V-8.

The fact is, I would *still* like to have one of these cars. No *Hemmings* or *Old Car Trader* goes through our house without my checking under "M" on the current price and supply of '49–'51 Mercs, right along with MGs, Morris Minors, Minis and Morgans. M is a car-rich letter of the alphabet.

But my scan of the Mercury section stops after 1951. No Mercury

made after that date seems to light my particular brand of kindling. The Cougars, Cyclones and Marauders of the 1960s and early 1970s were certainly noteworthy performers, but I never cared much for their styling.

Since then, there hasn't been even a flicker of interest. I have generally regarded Mercurys as alternate, restyled Fords with a few added amenities and extra trim. Conservative, comfortable and slightly bland.

Odd, I thought as I picked up my guitar again, that a major division of an American car company could go for so many years without catching the eye of an enthusiast (yers truly) who has a wide interest in all kinds of cars.

Would James Dean and Natalie Wood yearn to ride to school in any Mercury of the recent past? Would K.C. Douglas or Steve Miller win their girlfriends back again by buying a Mercury "to cruise up and down this road?" I didn't think so.

Things had become awfully quiet at the Mercury division in recent years, and it wasn't a brand one heard mentioned very often among friends who were shopping for cars. Was Mercury another great name destined to go the way of Plymouth and Oldsmobile, a redundancy in a world awash with other choices?

"I'd wanted a 1949 Merc ever since seeing Rebel Without a Cause when I was a kid."

As if to answer this very question, the phone rang just one day later. It was Editor-in-Chief Tom Bryant.

"Mercury is organizing a sort of 'Great American Road Trip,'" he told me, "taking the new Mercury Grand Marquis, which they point out is one of the last of the great American 6-passenger luxury cars with a V-8 in the front and rear drive, on a musical tour of the Midwest. They plan to start at a Chicago blues club, drive up to the Motown Museum in Detroit and then go down to the Rock and Roll Hall of Fame in Cleveland, taking mostly two-lane roads through small towns. If you are interested in going, you can call their PR manager, Jeremy Barnes."

My friends would later accuse me of designing this trip myself and suggesting it to Mercury, which is completely untrue. It was a simple case of entrapment.

So there I was, two weeks later, shaking hands in downtown Chicago with Jeremy Barnes, his associate Jay Joyer and meeting four other journalists who were going on the three-day trip. One of them was my old friend Denise McCluggage, a fellow music buff from way back.

That evening we went to a club called B.L.U.E.S., which is an authentic smoky blues bar full of young locals, rather than a contrived theme bar for busloads of tourists. Willie Kent and the Gents were playing, an excellent blues band whose CDs I actually own. Nice evening.

Suitably saturated with cigarette smoke, we slept off our cheap booze back at the hotel and when I woke up in the morning, everything I had was gone.

Just kidding. Everything was fine.

We emerged from the hotel in the morning, loaded our luggage into four waiting Grand Marquis and we headed toward Michigan, peeling off the I-road onto the two-lane Highway 12 for Detroit.

Denise and I co-drove a dark red LSE version, which comes with a handling package, floor shift and dual exhaust system that boosts the output of the 4.6-liter V-8 from 224 bhp to 239 bhp. It also has a shorter rear-end ratio (3.27 vs. the standard 2.73) for better acceleration.

The chassis, our Mercury guides told us, has been redesigned this year, with a stiffer frame, speed-sensitive rack-and-pinion steering (out with the old recirculating ball) and new suspension geometry on the Watt-linkage live rear axle.

Whatever they did, it works. On the road, the LSE is really quite nice. Quick, quiet, responsive and comfortable, with a nicely damped ride and surprising nimbleness and lack of body roll in corners. It also has the kind of natural steering accuracy and feel that only a front-engine/rear-driver ever seems to produce. Furthermore, the rubber-mounted separate body is free from even a hint of gnawing road and tire noise.

Essentially, I couldn't find anything to dislike. It was the perfect car for a trip on the open road. It even looks nice—maybe better than the Ford equivalent, for once.

Driving into Detroit, we cruised down West Grand Boulevard and pulled up in front of the Motown Historical Museum. This famous studio, founded by Berry Gordy Jr., started out in an old house, and as the company grew, Gordy simply bought more houses on the block and converted them to offices and rehearsal studios.

He had tryouts in an audition room one afternoon each week, and from this talent search emerged The Supremes, Smokey Robinson and the Miracles, Marvin Gaye, the Four Tops, Stevie Wonder, the Temptations, Martha and the Vandellas, etc., most coming from the homes and schools right near the studio.

Amazing what an urban neighborhood will yield if you merely ask.

After a night at Ford's famous old Dearborn Inn, we visited Ford World Headquarters to meet briefly with none other than Elena Ford, granddaughter of Henry and Brand Manager for Mercury.

For some reason, I had expected a debutante type in a cocktail dress who would have to set her champagne glass down for a moment to chat with us. Instead, Elena Ford turned out to be a modest, busy, no-nonsense woman in a business suit who has a quick, inquiring mind and a very accurate sense of what Mercury is—and isn't.

She said they are trying to revitalize the Mercury name and product line with cars like the revamped Grand Marquis and new Marauder, and to re-enter the American consciousness with livelier advertising, such as the new "Crazy About a Mercury" campaign with Dwight Yoakam and Aretha Franklin singing "Mercury Blues."

There was that song again.

Later that morning, we hit the back roads down to Cleveland, where we parked at the Rock and Roll Hall of Fame, a huge glass pyramid along the lake. The place has six levels joined by escalators and filled with Rock memorabilia, theaters and displays.

My favorite display, for some reason, was a small glass case containing the tiny, roughed-out suede cowboy boots worn by Michelle Phillips on the famous "bathtub" cover of the first Mamas and Papas album. They were beaten up and all worn out.

I looked at them for quite a while. There was something indefinably poignant in those boots, and I think it had to do with lost youth and the passage of time. Two of my biggest problems.

Late that afternoon, we all said our goodbyes and headed for the airport.

On the flight home from Cleveland I found myself sincerely wishing Mercury well in its effort to liven things up. The company has some perfectly nice cars in its lineup with the Mountaineer, Sable and Grand Marquis, but, until now, there hasn't been much to inspire a blues tune about jealousy and desire. Maybe the new Marauder will do it, or other designs still on the drawing board. I'd love to see it.

"Mercury Blues" is a very old song about a very old car. Maybe it's time we had some new cars enter our national mythology. It's getting harder to steal one's girl back from one's best friend using a car from 1949.

March 2003

My Brother's Cars

When the phone rang a few weeks ago, I instantly recognized the voice of my brother Brian, coming to me from Nashville through the quavering fog known as cellphone technology.

"Am I the last male in America," he asked without preamble, "who still wants a car rather than an SUV?"

"You must be car shopping," I said. "What happened to your Jaguar XK8?"

"I was driving to the airport the other day and a confused older person decided to come to a full stop in the middle of Hillsboro Pike. I was the last car in a six-car pileup."

"You okay?"

"Yeah, but the Jag may be totaled, so I'm looking at other cars. All the guys at work are telling me to get an SUV, and I've been out test-driving a bunch of them."

"What do you think?

"I don't get it. These things aren't any fun to drive! They handle like trucks and get worse gas mileage than my Jaguar. Why do people buy these things?"

I glanced nervously out the window at Barb's Jeep Grand Cherokee, which sat in our driveway, and cleared my throat.

"Well," I said, putting on my best Expert Witness voice, "people buy them because you can carry a canoe on the roof rack and tow a trailer, and because they carry four or five adults in comfort. They do everything most cars used to do, but don't anymore. Also, they go through deep snow."

There was a silent hum of thought on the other end, and then Brian said, "Maybe so, but they sure take the fun out of driving."

"Yes they do," I said, secretly glowing with older-brotherly pride. "You don't need an SUV. Get a car you like to drive."

"If they can't fix the Jag," he said, "I might just get another XK8. I love that thing, and it's been absolutely trouble-free, but it's still fun to look around at other cars...."

I've been getting car calls like this from my brother for many years. From many cities.

"I'd lean the 2-year-old Brian against a tombstone, give him a few graham crackers to keep him busy, then do timed laps around the cemetery road."

Brian has spent most of his life in radio, but is now general manager of Country Music Television, a job that allows him to work regularly with performers whose faces I know only from album covers.

Just a few months ago, he called me late from his car phone and said, "You'll never guess where I am right now."

"Probably not," I replied truthfully. Last time he posed a brain teaser like this he was at a party with the Rolling Stones.

"I am, at this moment," he said reverently, "driving my Jaguar out through the gates of the home that used to belong to George Jones and Tammy Wynette, and now belongs to Earl Scruggs and his wife. I've just spent the whole afternoon at Earl Scruggs's 76th birthday party, listening to the best bluegrass musicians in the world jam in his living room."

Brian knows I enjoy getting these calls because we have almost exactly the same taste in music, not to mention cars, books, movies, etc. If he appreciates something, he knows I probably will too.

Which is odd, in a way, because I was 13 when Brian was born, and we didn't exactly grow up together. Our similar musical tastes, however, can be partly explained.

When I was in Vietnam, my brother was only 8 years old, but he more or less took over my record collection. He would sit in his room for hours, rocking in a child's rocking chair, listening to the Stones, Beatles, Dylan, The Band, Paul Butterfield, etc. In other words, his small, developing brain was ruined forever.

Other brain damage probably occurred because I used to take him riding on my go kart out to our local cemetery when I was supposed to be "babysitting." I'd lean the 2-year-old Brian against a tombstone, give him a few graham crackers to keep him busy, then do timed laps around the cemetery road.

It was sort of an Edgar-Allan-Poe-meets-Steve-McQueen environment. Between the crumbling old tombstones and ear-splitting noise, it's a wonder Brian didn't turn into a car-hating Goth.

But he didn't, and I proudly observed from a distance that his very first car was an MG Midget. Brian bought the car when he got his first radio job at a small Country radio station in rural Florida. He loved the Midget, even though it broke down nearly everywhere he went.

Then came the first of the car-related phone calls.

"I sold my Midget," Brian said with wonderment in his voice, "and bought a Datsun B210. I've been driving for over a week and nothing has gone wrong! I can go anywhere!"

And nothing did go wrong with the Datsun, until he rolled it while crossing the Everglades on Alligator Alley one night.

Brian swerved to avoid a cat and lost control. With no seatbelt on, he was thrown through the windshield and landed in the middle of the road on his hands and knees. The car catapulted over his head and disappeared neatly beneath the swamp water in a roadside ditch.

Brian stood up, bruised and cut, and looked around. No car. He walked toward the lights of a nearby house, where a man on the porch threatened to blow his head off with a shotgun.

So Brian wisely walked to the next house, farther down the road. There the people called the police, who thought Brian was on drugs, because they couldn't find his mythical car. It was finally located when someone noticed a smashed Allman Brothers cassette and paint marks at the edge of the road. The lily pads had closed over the car perfectly, leaving not a scar or a ripple, like Sandburg's *Grass* at Waterloo.

Brian had been speeding home to see our mother, who was gravely ill at the Boca Raton hospital, and he'd been told to come right away. He caught a bus after the accident and arrived home with dried blood on his knees.

I got the same call in Wisconsin and flew down that night. After visiting our mother, we left her to rest and I drove Brian in my rental car back to a wrecking yard on Alligator Alley so he could retrieve what was left of his luggage.

When I saw that smashed, windowless car full of mud and weeds I got tears in my eyes. The roof had been flattened down over the steering wheel.

My dad later said Brian's survival that night was nothing but a simple miracle, "because God knew we couldn't take any more."

Since then, I've gotten many other car-related calls from Brian, all of them in happier times.

"I just test-drove a new Firebird Trans-Am," he told me in 1984, "and loved it. What do you think?"

"Great car," I said. "I'd buy one myself, if I weren't racing a Formula Ford."

Brian enjoyed the Firebird for several years, but started to have repair trouble as the miles wore on. He called again.

"Just test-drove a Porsche 944," he said. "Seems like a really fun car, tight and nimble. What do you think?"

"Great car," I said. "I'd buy one myself if I weren't restoring a Lotus Seven."

A few years later he called from a street corner pay phone in Philadelphia and said, "I went to trade my 944 in on a newer one and the salesman was so snooty I went across the street to the Jaguar dealership."

"And?"

"And they loaned me a black XJ-S convertible with a V-12 and said I could keep it overnight. I think I'm in love. This thing has wonderful suspension."

"Great car," I said. "but you're going to have more car trouble than you did with the 944."

After a brief silence, Brian said, "I don't care. The car's so nice to drive, it seems worth some extra trouble."

My life story, in one succinct line. And maybe his.

A few years ago he called to get my thoughts on a British Racing Green Jaguar XK8 roadster with a tan interior. Might as well ask Jacques Cousteau how he feels about seawater.

"My favorite current GT car," I said.

Then came the recent Jaguar accident on Hillsboro Pike. What would come next?

I rode my motorcycle from Wisconsin down to Nashville last week to visit Brian and meet his fiancée, Sarena. When I arrived, there was a brand-new black Corvette in the driveway.

He told me the XK8 would soon come home from the repair shop, too, but he couldn't resist the Corvette. "This is a great car," he said.

"Wouldn't mind having one myself," I said, "if I weren't restoring an E-Type coupe and a Lola 204."

And I guess that's always been the main difference between us.

I was born mechanically inclined and Brian *electro*-mechanically inclined. Brian can run a sound studio full of electronic gadgetry but he can't put the top back on a ketchup bottle straight. (The threads are too small.) I can rebuild a Jaguar, but can't program a digital alarm clock. He works late in the studio so he can own nice cars, and I work late in the garage so I can redeem nice cars. But in spirit and aesthetics, they are essentially the same cars.

And once behind the wheel, we both turn up the radio when the Rolling Stones or Hank Williams come on. Just born that way, I guess.

April 2003

Charts of the Ancient World

My old friend and neighbor Chris Beebe stopped by the other day and dropped off a big, flat plastic shopping bag.

"I thought you might enjoy this," he said with his car idling in the background, a sure sign he's about an hour late for something and can't talk very long. Chris owns a repair shop called Foreign Car Specialists, and, like most busy people, lives with the certainty that to please one person is almost always to disappoint someone else.

I reached into the bag and pulled out a large State Farm Road Atlas of North America, printed by Rand McNally.

"Looks old," I said.

"It's from 1954," Chris said. "The era just before Interstates. Thought you might have fun looking at it."

"Where did you get it?"

"The service manager at the shop, Jerry Wideen, found it at a garage sale, and I traded him a disassembled metal building for it."

I knew it would take too long to unravel the Byzantine workings of a barter economy in which a 49-year-old road atlas might be traded for a collapsed metal shed, so I thanked Chris and took the atlas in the house. Coffee in hand, I sat down to examine it.

Ah yes, 1954.

I would have been 6 years old when this atlas was printed, and my mom and dad used one exactly like it on long road trips. The 1950s' cover art had a look of instant familiarity, like a movie poster for a Gene Kelly musical, preferably with Leslie Caron or Cyd Charisse.

A colorful map of the U.S. was scattered with cartoonish figures intended to show the attractions of each state. Florida had palm trees, a pink flamingo and a sailfish leaping offshore; Alabama and Georgia had happy black people sitting on cotton bales and playing banjos (which, I'm sure, is exactly how James Baldwin or Eldridge Cleaver remembered the 1950s); California had crossed palm trees, a movie director in a beret behind a camera on a tripod, and some grapes and redwoods in the north, and so on.

My home state, Wisconsin, was dominated by a huge beer stein, a Guernsey cow, a milk can, a leaping deer and some pine trees in the north. Which pretty much sums up the reasons I'm here, if you factor in Road America.

The only really puzzling sketch is Utah's. That state is symbolized by a vegetable that looks like a beet, roots, leaves and all. Maybe Utah used to be "The Beet State," and I never heard about it.

The map shows its age by having Nevada's symbol a sun-bleached longhorn skull and a cactus. No slot machines or roulette wheels. But this was probably early in the gambling era, when Las Vegas was not yet a national treasure. A quick flip to the city listings in the atlas shows the population then was just 24,624, based no doubt on the 1950 census.

I got out my new 2003 Rand McNally and found the current head count of Las Vegas to be 478,434. One of whom is probably Wayne Newton.

"You picture these cities of old rising abruptly, like the towers of Oz, out of the farm fields and forests of the surrounding countryside."

Speaking of population, I checked my own hometown of Elroy, Wisconsin, and found it listed in 1954 as a city of 1654 souls. And that's about how big it was when my family moved there from Minnesota in 1952. By the time I was in high school, however, the population had dropped to 1503, and it stayed that way for years. It was a railroad town, a major switching yard for the Chicago Northwestern, and the railroad had nearly faded away by the mid-1960s. The population now is 1578.

Looking at the old atlas, the roads in and out of Elroy haven't changed much, except to be straightened and made safer. For instance, the steep and tortuous Good-Enough Hill between Elroy and Mauston used to kill about three people a year, by way of drink or snow. (And I once survived the trip with a guy who was wrestling, simultaneously, with both of these devils.) The old road was finally bypassed—making the drive easier but infinitely duller—just before I got my driver's license. Our town lost a little of that comfortably remote, snowed-in feeling.

As you get farther from town, however, the absence of Interstates gives the map a pioneer-era quaintness. You can see that the current Interstates—I-90 and 94—either parallel the old two-lane state highways or were built on top of them, as with the now mostly-obliterated Highway 30 from Madison to Milwaukee, but the map has no big, straight highways that shrug off the natural terrain. The old roads all look more like rivers than pipelines, and you get the feeling not much property was condemned to make them. In 1954, the only four-lane stretch of road in all Wisconsin was Highway 41 linking Milwaukee and Chicago.

The speed limit chart in the back of the atlas, incidentally, shows the Wisconsin highway limits to be 65 daytime and 55 at night. (Now it's 55 on two-lane roads and 65 on the Interstate.) And nine states still list the upper limit for daytime as "Reasonable and Proper." Those were the days.

It's interesting to contemplate my parents' chosen routes during the '50s from Elroy to Florida in the winter. One year we took Highway 14 down to Highway 13 (now 213), getting stuck in a blizzard at a place called Magnolia, just 15 miles from where Barb and I now live.

Every time we drive through Magnolia these days, I always say, "Here's Magnolia Hill, where we got stuck in the blizzard in our old Buick!"

And Barb always says, "I know," and pats me on the knee. If I tell her this one more time, I'll need knee patches on my pants. Or football pads and a helmet.

That year, as always, we drove right through the middle of downtown Chicago's Lakeshore Drive on Highway 14 and then somehow ended up on Highway 231 all the way to Dothan, Alabama, cruising into Florida on Highway 27. Four days on twisting two-lane roads, right through the middle of every small town in four states, following trucks, avoiding drunks and taking endless detours for road repair. It made those palm trees, pink flamingos and leaping sailfish look mighty good when you got there.

"The old roads all look more like rivers than pipelines."

Trips east, to see my grandparents in New York, were a little better. There were a few sections of four-lane road on Highways 20 and 40 in Ohio, and then you hit that godsend, the marvelous Pennsylvania Turnpike. This early super highway must have been built purely as a humanitarian project to prevent countless fatalities in this rugged state, with its fogs and frozen rains. Even on the Turnpike, I have vivid memories of several terrible accidents we passed, where semis had slid right through guardrails and lodged in the trees below.

Going west, to see our other grandparents in California, there were virtually no four-lane roads until you got to Sacramento in Northern California or Palm Springs on your approach to Los Angeles, but at least the highways were fairly straight. Route 66 still appears on the old atlas in nearly all its unmolested, Bobby-Troup-era glory, winding from Chicago to L.A., more than 2000 miles all the way. The Turner Turnpike, however, does bypass a section between Tulsa and Oklahoma City, and there are a few short four-lane bits around St. Louis and between Bloomington and Springfield, Illinois.

Los Angeles doesn't really look very different on the old map. In 1954, the L.A. basin was already laced with four-lane boulevards or

"Super Highways" that foreshadow pretty accurately the network of current freeways. Same with New York. Big cities had lots of traffic to handle, even then.

Although I see Newport Beach, home of R&T, had only 12,120 people then and it has 70,032 now. No wonder we can't test cars for top speed on Brookhurst Street anymore.

But what you notice most—and perhaps most nostalgically—on the old road atlas is how quickly suburban sprawl drops off at the edges of the big cities. Denver, on this map, looks like a compact little downtown surrounded by a few residential neighborhoods. You picture these cities of old rising abruptly, like the towers of Oz, out of the farm fields and forests of the surrounding countryside, unannounced by miles of shopping malls, car dealerships and fast food joints.

This is a vision of cities I still prefer—like those illustrations in the old flight training manuals I had in high school, where a metropolis is represented by a small cluster of smokestacks and skyscrapers on the agrarian horizon—but it is obviously not the one most Americans prefer, or we wouldn't have changed it so radically.

It was unavoidable, I guess. My new World Almanac says the population of the U.S. in 1950 was 151,325,798, and in 2000 it was 281,421,906. That's a big increase, and, as the saying goes, everybody's gotta be somewhere.

And I seem to have ended up, once again, just south of the beer stein and slightly east of the Guernsey, within driving distance of the leaping deer and the pine trees, where the smaller towns still have edges, and the roads follow the landscape like rivers. Except for lower speed limits—and having a streamlined, high-tech 1967 Jag in my garage—it could still be 1954.

May 2003

Driver's Ed.

Yesterday's headlines were not so good. Three young people were killed, a few miles from where we live, when their much-modified Honda Civic went through a stop sign and straight off a T-intersection into a field at very high speed. Police said "alcohol was a factor."

Only one ray of good news shone through the dark cloud of this disaster. A fourth passenger, a high school girl, saw that the car was about to catapult off the road and had the presence of mind to reach over and click her seatbelt into its socket. She escaped with minor injuries, while the three young men not wearing belts were fatally thrown from the car as it rolled.

"Here we go again," I mumbled, reading the story. It was at least the third time in the past month that young people in our area have been killed while not wearing seatbelts. In fact, we've had such a rash of these accidents lately that the local newspapers are beginning to run editorials asking what's going on. A good question. According to the paper, a recent U.S. Department of Transportation survey says that 67 percent of all drivers use seatbelts, but only 55 percent of the youngest drivers use them.

Surely their driver's education programs have explained the simple physics of car accidents and shown the usual films of crash test dummies being thrown violently around the insides of crashing cars. The lesson is basic and clear, so why no belts?

Peer pressure?

Probably. Insecure kids fear appearing uncool more than death itself, and in some circles it's considered a sign of weakness to worry about safety. Careless fatalism has ever been hip among the young.

Or it might be simple laziness, or even the persistence of that tired old theory that "it's better to be thrown from the car."

On that last heresy I must admit some complicity. I wrote a column recently about my brother surviving an accident years ago in which he was miraculously thrown through the missing windshield of a car (it had popped out during the end-over-end tumble) just before the car sank beneath a swamp next to the highway. He walked away, almost uninjured.

I hesitated even to mention this accident because it was virtually the only case I've ever known in which a person was better off being thrown from a car. And I say "virtually" only because a few race drivers of the distant past, such as Masten Gregory and our old friend Innes Ireland, survived crashes by bailing—or catapulting—out of their fragile old "flying gas tank" race cars.

"Mr. Ray was my driver's education teacher in high school.... he called me out of study hall to go driving with him one winter day late in 1963, and his first words to me were, 'They tell me you're a wise-ass. Is that true?'"

But in the modern era of chassis design, one is clearly better off restrained within the chassis rather than slamming into it or sailing out of it. Seatbelts have probably saved my own life twice—once on the street and at least once on the track—during extremely rapid stops in which I bent the steering wheel with my hands but stayed in my seat. I'm a believer.

But then I was a confirmed seatbelt user well before these accidents occurred or I probably wouldn't be here to talk about it. Why should this be?

Well, it's certainly not a natural infatuation with safety, because I tend to have the kinds of hobbies that make insurance agents crazy, and I've driven all kinds of old cars without belts in recent years, giving the added risk no more than a passing thought. (Yes, even at this age, I still enjoy those moments of careless fatalism I mentioned, without which there is not much levity or adventure on this earth.)

But if a car *has* belts, I am almost physically incapable of putting the car in gear without first reaching for a belt and hearing that click of metal against metal. It's an automatic ritual, like shouting "Clear!" out the window of an airplane before spinning the prop. Can't help myself.

Taking off in a car without the seatbelt on is like dressing with one shoe, or leaving your billfold at home. A dim mental alarm bell (and I have lots of those) goes off, warning subliminally that something's wrong.

So how, you ask, did a devil-may-care, heedless, swashbuckling and absent-minded young rake such as I become such a dutiful and robotic user of seatbelts?

Well, I think Mr. Ray is to blame.

Mr. Ray was my driver's education teacher in high school.

He was a big, burly guy with a good sense of humor, and I took an immediate liking to him. He called me out of study hall to go driving with him one winter day late in 1963, and his first words to me were, "They tell me you're a wise-ass. Is that true?"

I stared at him in confusion for a minute because no teacher had ever been so blunt—or so accurate—before, and then I broke out laughing. "Yes," I said, "I guess it is."

"Well, let's go driving."

As one of the oldest kids in my class, I was the first chosen for this sacred instruction upon which all life, glory and social progress depended. He also called out the name of Mary Ellen Leatherberry, so Mary Ellen and I followed Mr. Ray out through the snow-packed, windswept parking lot of Royall (sic) High School in Elroy, Wisconsin, and climbed into a brand-new metallic turquoise 1964 Ford Galaxie, on loan to the school from Northcott Motors.

Over the next few months Mary Ellen and I trundled at precise speed limits up and down the steep hills of Elroy, parallel parked on Main Street over and over again and practiced handbrake-release starts on hills. We also chauffeured Mr. Ray on all kinds of official school errands—to the bank, to the bakery to pick up donuts for the teachers' lounge and to the grocery store for cartons of cigarettes. This was an era when nearly all teachers smoked, "to steady their nerves," and the teachers' lounge appeared to be perpetually on fire, like an aircraft carrier with battle damage.

And, Lord knows, Mr. Ray needed to steady his nerves more

than most.

Yet he always seemed calm and at ease, and, during those many hours of driving around town, Mr. Ray somehow managed to tattoo upon our very souls three basic habits: (1) The key was never turned in the ignition until all the seatbelts were on; (2) turn signals were used for every lane change, turn or anticipated lateral movement of the car; and (3) we never backed up so much as an inch without turning around fully in the seat and looking behind ourselves. No backing up anywhere using the mirrors.

All of these things have stayed with me.

Just as I'm nearly incapable of driving without hooking up a seatbelt, I also have a Pavlovian need to use the turn signals in concert with the steering wheel when I turn. I don't think I could make a normal, non-emergency lane change without turn signals, even at gunpoint. It's just a built-in part of the maneuver.

Almost comically so. I've raced showroom stock cars a few times and have actually found myself signaling a lane change on the front straight or during a sudden pass under demon late braking. Even on a back country road, I sometimes "accidentally" signal for a sharp turn when there is no stop sign and nowhere else I could go. I'm just a signalin' fool. Mr. Ray did this to me.

So when I see other motorists suddenly change lanes without signaling, I am both offended at their lack of manners and amazed that they can do it at all. To turn without signaling seems, to me, almost as hard as not returning a friendly wave; it requires either a straitjacket or a high level of grouchiness.

Turning around to back up is another matter. I've been backsliding here a bit. When my lower back was in better shape and my neck more youthfully limber, I always turned fully around to back up. A good thing, too, as it saved me many collisions with objects I would not have seen. Now I cheat a little and sometimes use the mirrors. Which is how I backed over that flower planter at the Union 76 station in Newport Beach a few years ago.

As I was helping the station owner scoop the dirt back into the fallen planter, I actually thought of my old driver's ed. instructor and was glad he couldn't see me now. Or maybe he could. I lost track of Mr. Ray after high school, and he moved away from town about the time I left for college, so I don't even know if he's still alive and well.

But I am. Thanks to him, and a few pieces of nylon webbing.

June 2003

The Two-Car Time Machine

So there I was the other night, standing in our newly expanded Borders books and music mega-store, holding two CDs and trying to decide which one to buy.

In my right hand was a Mitch Ryder and the Detroit Wheels anthology (to replace my badly scratched vinyl original) and in my left hand was a newly repackaged Maria Callas compilation of great arias. Yes, these two unlikely combatants were fighting for my soul. And losing.

Why?

Well, because one CD was priced at $17.99 and the other was $16.99, despite production costs having been amortized about 30 years ago, when this same music was available on vinyl for about $2.99. And the record industry wonders why it's in trouble.

Market Place

FOR SALE

"1953" MG-TD Sport Roadster. 492 actual miles. Heater, Jet black finish. New car title and guarantee. Sacrifice $1845. Bank Rate Terms. Glenn Walraven, Marion, Ohio. 2-4135. Nights 2-5708.

FERRARI—Ex-Bill Spear 4100 cc Competition Roadster Superleggera Body by TOURING. Fresh overhaul, New Crank, Bearings, Rings, Valve Job, New Gearbox. Multiple Disc Racing Clutch. Extra Set Wheels. Can furnish Custom Trailer. Asking $7,000. Charles E. Brown, West Monroe, Louisiana.

1951 FERRARI AMERICA 4100 cc Vignale two-place coupe, black and green, green leather upholstery, engine recently overhauled at Modena, 250 hp, phenomenal performance, new gearbox, $7500. V. Corradini, 445-60th Street, West New York, N. J.

ROAD and TRACK, January, 1954

So I put both CDs back in the racks and wandered toward the automotive section of the bookstore, which is now stuck back in a remote corner, next to Science and Physics, pushed there no doubt by the ever-encroaching glaciers of Self-Help and New Age books.

The concept of New Age *books*, incidentally, seems odd to me. You'd think these ideas would be better spread in an oral storytelling tradition, by hooded figures gathered around a large bubbling cauldron of newt's eyes and bat's wings, but what do I know? Not much, but at least I'm easily distracted.

While headed toward the car shelves, I glanced at a display of new science books and a green paperback caught my eye. It was called *Time Travel in Einstein's Universe* by J. Richard Gott. The back cover explained that Gott is a distinguished professor of astrophysics at Princeton who writes for *The Scientific American*, *American Scientist* and several other esteemed publications. A smart guy, in other words. No New Age nonsense here.

I was naturally attracted to this book on time travel because—like all car buffs—I have always dreamed of having a time machine about the size of a two-car garage that can be catapulted back into the past.

The advantages to such a device are obvious. You could buy, say, a couple of new 289 Cobras from a Ford dealership at their original $5195 list price. Or a slightly used D-Type Jaguar or Ferrari GTO for even less. Or just a Series I E-Type coupe like mine that doesn't need three years of knuckle-busting restoration before you can drive it. Better to drive a new one right out of the showroom.

You get the picture. All you have to do is look in the R&T classifieds from the '50s and '60s to see lost opportunities that make your heart stop. How wonderful to go back there in your time machine/garage, pick up a few of the world's greatest cars for a fraction

JAGUAR XK-120 cpe w/Buick V-8 engine, all new equipmt including int. and beautiful candied maroon paint. Mint cond. Sacrifice $970. Anne Taylor, 5318 Via Del Valle, Torrance, Calif. FR 5-0386.

1951 FERRARI Mexico Vignale cpe. Red w/black leather upholstery. Almost new Corvette engine and 4-spd box. Asking $4500, no trades. D. A. Hernandez, 936 White Dr., Santa Clara, Calif. CH 3-2442 or CH 1-2051.

JANUARY 1962

of their current value and bring them home to the present.

Obviously, this wouldn't be as simple as it sounds. There are complications.

First, the transfer of money. It wouldn't do to show up in the past with a suitcase full of money printed in 2003, or any time during the recent past. One look at those "large cerebrum" Andy Jacksons we have now, and the potential seller might alert the cops. I don't know how one would go about locating genuine old money, short of finding it stuffed in some old geezer's mattress. You'd have to advertise. Or use gold, and accept some rate-of-exchange losses.

The only other headache is finding appropriate attire. This is no problem for me, as I still have the suit and skinny tie I wore to the

JANUARY 1968

1946 MG TC. New engine, BRG paint, top, tonneau. Rebuilt transmission, front end, speedometer. Seat leather cracked. 19-in. wheels (original). $2250 firm. For appointment, write: J. Frey, 812 Portland Terr., Alhambra, Calif.

1955 XK-140 roadster. Exquisite hand-rubbed BRG lacquer, new black interior. 3000 miles since total rebuild. New: engine, clutch, top, Pirelli HS, Abarth, wiring, brakes. $3500. T. Booth, 1412 N. 45th, Seattle, Wash.

LAMBORGHINI 400GT 2+2 coupe, one year old. Body by Touring Superleggera. 8000 miles under strict factory control and service. Brilliant black with natural beige leather interior, chromed wire wheels, fully equipped, power windows, 4 power disc brakes, AM/FM signal-seeking radio, custom made specifications, electric chronometer, twin headrests, special low bottom seats. No accidents, never raced, absolutely new throughout. $9500, delivery anywhere. Fred A. Pettinella, 29 Washington Square W., New York, N.Y. 10011. (212) 982-8191.

homecoming dance in 1966, but you have to be careful here, not just with clothing, but also with accessories. It wouldn't do, for instance, if you were negotiating to buy a new Porsche 550 Spyder from Competition Motors (before James Dean had a chance to snap it up) and the cellphone suddenly went off in your pocket.

Also, body piercing would be out of the question. Grown men and women with tongue and nose studs were not as widely admired during the '50s or '60s as they are now. Nix on the heavy tattoos, also. When I was a kid, the only non-felons who had tattoos were globe-trotting Navy men or veterans of elite combat units. No sense souring a deal on a used $1500 Bugatti Type 35B because you have dragons crawling up your neck.

But these are minor problems, compared with the actual physics of time travel.

Or so I learned when I took Gott's book home and sat down with a big cup of coffee to read it that evening. And for several evenings thereafter.

The book starts quite charmingly with a survey of famous books and movies about time travel—*The Time Machine, Back to the Future, Somewhere in Time* and so on—exploring the internal logic and consistency in each story. Then it gets into a history of pioneering work in physics by Maxwell, Roemer and others, with a layman's description of Michelson and Morley's experiments on the speed of light and the observations that led Einstein toward his theory of special relativity.

"While I trust the future to improve the efficiency and practicality of cars, I don't trust it to make a car that is significantly more charismatic or beautiful than the Ferrari GTO."

Gott also points out some of the paradoxes of time travel itself, of which the most fascinating to me is the problem of the "jinn." This is some object or collection of particles that go around and around in time, without apparent beginning or end. An example is the pocket watch in the movie *Somewhere in Time*. An old woman gives Christopher Reeve a watch and he goes back in time, falls in love with her younger self and gives it to her as a gift, just before he's accidentally transported back to the present. So where did the watch come from? Who made it? An interesting problem, and one whose ramifications I can almost grasp.

So far, so good.

But as the book goes on, the author gets into quantum mechanics, cosmic string theory, wormholes, models of an oscillating universe and other concepts that represent a leap of brilliance and intuition that my poor brain cannot begin to make. He might as well try to explain the principles of flight to a bucket of Kentucky Fried Chicken.

But then physics was never my strong suit. As a junior in college, I suddenly switched my major from Journalism to Pre-Med and had to take a 5-credit physics course from a soft-spoken professor with a nearly unintelligible Chinese accent. The next thing I knew, I was flying around Vietnam in a helicopter.

A chain of events, you will note, more closely related to Darwin than Einstein.

So Gott's book is a little deep for a person of my limited wattage, but I was able to gather that time travel to the future—given the right booster rockets—is currently more feasible than a return to the past. He doesn't rule out either possibility, however, noting that we now live in a world of technological achievement almost unimaginable 100 years ago. We went from doubting the possibility of heavier-than-air flight to landing on the moon in less than the span of a lifetime, so why not time travel?

Gott doesn't see this happening anytime real soon, but, if it does, I am more likely to sign up for a trip to the past than the future. While I trust the future to improve the efficiency and practicality of cars, I don't trust it to make a car that is significantly more charismatic or beautiful than the Ferrari GTO I got to drive a few years ago.

So, given my druthers, I'd just as soon go back in time and search for that cast-off GTO Innes Ireland didn't buy at the end of his 1964 racing season because $5000 was just too much for a used-up old Ferrari GT car.

There is, of course, another advantage to visiting the past with a time-traveling garage, and it just occurred to me the other night while I was polishing the cam covers on my E-Type.

Cars could be transported both ways.

I could take the E-Type back and give it to myself in high school.

Think of it: The Jag would make a great car for dates, and it would be a lot more fun to drive than my dad's Falcon wagon. Plus I'd have the use of it all those years, right up to the present. The ultimate jinn!

The only problem with this plan is that it might be hard to explain to my parents how I suddenly acquired an E-Type coupe with my lawn-mowing money.

I guess I could tell them it was a gift from a mysterious, kindly older gentleman with a gray beard and scars on his knuckles. And a new Mitch Ryder album tucked under one arm.

July 2003

Random Reflections at 55

Feeling reasonably festive, I woke up on my birthday early this morning and went to the window to see what kind of day it was.

Hard to tell, really, because the glass was paved over with a solid sheet of frost. Weak winter light glowed dimly through finely etched crystalline patterns that looked like white pigeon feathers.

It was quite beautiful, but I must admit it jars the senses to roll up a window shade and discover a wall of ice three inches from your face. Sort of like opening your shower curtain and finding Norman Bates, dressed as his mother. Not exactly what you had hoped for, and a bit disappointing.

"Should a man really have to rebuild rusty old brake calipers on his birthday? No, he should not ...."

But then disappointing weather has always been a small part of every midwinter Wisconsin birthday. When I was a kid I was always getting birthday gifts intended for use in warmer seasons—baseball glove, fishing reel, Boy Scout knapsack, gas-powered model airplane, etc. Summer stuff.

One year I got a Winchester .22 rifle and made my poor dad drive me out in the country in howling subzero winds just after a blizzard so we could go target shooting. I shot tin cans off the tops of fence posts, until snow drifted over the fence line and there was no place to put the cans. We looked like a couple of Russian soldiers defending Stalingrad.

Nevertheless, the Winchester was a nice gift, and it's still hanging on the wall, exactly 43 years later. I glanced at it as I went downstairs, where Barb was making me a hearty birthday breakfast of waffles, bacon and coffee. She said the temperature outside was zero.

"That's not very many," I observed.

After breakfast, I decided to treat myself further by avoiding "real" work for the morning and I went out to the garage to work on my race car—speaking of toys you can't use until spring.

"A car that looks bad in photos almost never looks beautiful in 'real life.'"

There I had intended to finish rebuilding the rusty old Girling brake calipers on the Lola 204 Formula Ford project car I bought a few years ago. The pistons, of course, were stuck in the calipers like King Arthur's sword in the anvil, so I'd soaked them overnight in some stinky penetrating oil and now had to force them out with compressed air.

This is a messy, slightly dangerous business that usually fills the air with a fine mist of brake fluid, which is second only to mustard gas as a wonderful thing to inhale. I put on my green coveralls and stood at the workbench, staring ruefully at the tray full of cruddy brake parts.

Yet, inexplicably, I did not begin working.

How many sets of Girling or Lockheed calipers had I rebuilt in my life? A dozen? Two dozen?

Let's see: The first pair I did was on my Triumph TR-3, in 1968. After that, I did two more Triumphs, six Bugeye Sprites with disc brake conversions, three Lotus Sevens, two MGBs, two Jaguars, and four Formula Fords. And here I was 35 years later, doing exactly the same job on another British car. And my second Lola 204, at that.

Suddenly I felt very weary. Should a man really have to rebuild rusty old brake calipers on his birthday?

"No, he should not," I said to myself.

So I took my coveralls off and headed back to the house.

As it was too cold to do anything useful—or even useless—outside, I decided it would be a good day to kick up the heat in the furnace, drink a lot of hot coffee, get wired and wax philosophical about some aspect of turning 55. It was, after all, one of those minor milestone birthdays when a person should theoretically take stock. I would not work on cars this day, but rather reflect upon them. A clean, odor-free endeavor if I ever saw one.

I somehow resisted the impulse to lean on the fireplace mantle with one elbow and smoke a pipe, but I did manage to find a large yellow legal pad, a pen and a comfortable place to sit with a yellow cat on my knee doing a Sphinx imitation.

At the top of the legal pad, I wrote "Random Automotive Reflections at 55," followed by the list below:

1) Without British cars, my lifetime interest in automobiles and racing would have been reduced in its scope and intensity by considerably more than half. Yet I know all of their faults. Maybe there is such a thing as the Druidic Gene. My ancestors, after all, worshiped trees.
2) Most of the jokes about Lucas electrics would be better aimed at SU electric fuel pumps, which have caused more sudden roadside stops than the state highway patrol.
3) A car that looks bad in photos almost never looks beautiful in "real life."
4) Air cooling in car engines creates far more complication than is eliminated by the loss of a radiator, hoses and water pump.
5) Electrical engineers perfected interior car controls and switches about 30 years ago. Now, with nothing left to do, they simply complicate and bedevil our lives.
6) Go karts are every bit as fun and instructive to drive as race cars, at about a tenth the cost. All they are missing is that beautiful view down the hood and the illusion of glory. Neither of which should be discounted.
7) The advantages of mid-engine placement in a road car are largely theoretical, and there are many disadvantages. After all these years, front-engine/rear-drive cars are still my favorites. Except for the Mini. And the Boxster and the Ferrari 250LM.... Okay, and the 246 Dino and Renault-Alpine A110.... Never mind. Forget I said anything.
8) The "CHECK ENGINE" light on your dashboard could easily be

replaced by a simple flashing dollar sign. Every time mine blinks, it costs me exactly $450.

9) Art and technical perfection are not the same thing in cars, and hard-core car buffs often prefer art. Which is why they own so many tools.

10) The same Baby Boomers who ridiculed their parents for driving around in "2-ton land yachts" during the 1960s are now driving around in 2.5-ton SUVs that get the same dismal mileage, but have less luggage space. I oughtta know.

11) Cars have gotten continually "better" but no more fun, beautiful or interesting since about 1968. The same is true of the music emanating from their radios.

"Art and technical perfection are not the same thing in cars, and hard-core car buffs often prefer art. Which is why they own so many tools."

12) A news magazine reported last year that there were at least 10 national ad campaigns using the theme "No Boundaries," "No Limits" or "No Rules," many of them for either cars or restaurants. This year I think there are even more. These are odd concepts in a world troubled by ugly chaos.

13) Whenever I buy really expensive synthetic oil, I always secretly wonder if I will live long enough to reap its benefits.

14) "Ah, but when everything is working right..." is a preamble nearly all of us use when heaping praise on our British cars.

15) A normal-size adult in a Cadillac Escalade always appears to have his or her skull shrunken by the head-hunters of Borneo. It's a problem of scale.

16) Cars are said to be an art form, yet the Mona Lisa, I've noticed, never needs a cooling system flush or new brake pads.

17) The repeated, arbitrary yellow flags of Winston Cup racing, intended to increase viewer interest by bunching up the field, usu-

ally cause me to turn off the TV and go for a walk. Why not trip the leader in the Boston Marathon so the others can catch up?

18) The Germans, for my tastes, are currently winning the sedan styling wars.

"'Ah, but when everything is working right...' is a preamble nearly all of us use when heaping praise on our British cars."

19) Old Ferraris are always both far better and much worse to drive than you expect. The harder you drive, the better they are.

20) Racing an open-wheeled formula car is just a form of heroin addiction for people who don't like needles.

21) The original Issigonis-designed Mini is more fun to drive than any car I've ever been in. When everything is working right.

"The repeated, arbitrary yellow flags of Winston Cup racing, intended to increase viewer interest by bunching up the field, usually cause me to turn off the TV and go for a walk. Why not trip the leader in the Boston Marathon so the others can catch up?"

22) Road-racing circuits are meant to be bordered by green grass and shade trees. The only exception is Monaco. In racing, as in real estate, location is everything.

23) Straight-6s are the sweetest-running engines on earth, in my opinion, followed closely thereafter by V-12s and low-compression straight-8s. But then I've never driven a V-16, so I may be wrong.

24) Classic collector cars that irritate you in some fundamental way should come with onboard battery chargers, because they will soon need them.
25) Restoring a car yourself makes you feel like a doctor of orthopedics with x-ray vision. While others admire the patient's hat, all you can see is that badly healed fracture in the left clavicle.
26) People who have rebuilt more than a dozen sets of Girling brake calipers by the time they turn 55 should probably quit wondering why they are doing it, and simply enjoy the process. Then have some cake and ice cream and wait for spring.

August 2003

The Great Mini-Blitz

"Have you driven the new Mini Cooper S?" my friend Mike Mosiman asked over the phone late last autumn.

"No," I replied.

"Oh, man! You gotta drive this thing! I just bought one last week in gray and white, and I absolutely love it. I'll bring it right over so you can take a test drive."

That promise would have been quite reasonable if Mike lived in a nearby town or neighborhood. But he doesn't.

He lives in Fort Collins, Colorado, and I live in southern Wisconsin.

Nevertheless, Mike showed up at our door about 20 minutes later. Okay, that's a slight exaggeration; it was actually about two days later, but it seemed as though I'd just hung up and there was a Mini idling in our driveway. This is not the first time this sort of thing has happened.

Last summer, Mike rode out on his new motorcycle, a BMW R1150RT, so I could take *that* for a test ride. Unlike me, Mike tends to buy fast, reliable vehicles that are undaunted by vast distances. I buy stuff that blows up on the way to the mailbox.

Naturally, the minute Mike arrived we took the Mini for a long drive in the country.

The first thing that struck me about this supercharged car was not only that it was very quick, but that it was deceptively fast, cruising effortlessly at a relaxed and quiet 75–85 mph. Not at all the hyper wind-up toy I'd been expecting.

"Jeez," I said, "if I owned this car I'd get tickets all the time. We're going 80 and it feels like about 54 mph."

At that moment we crested a rise with the wheels practically off the ground and flew past a dark blue Ford Crown Victoria with an external spotlight, a whip antenna and writing on the door.

I stood on the brakes and looked in the mirror. The Crown Vic stood on his brakes too.

"Oh, brother!" I shouted, eschewing the sort of expletive I might normally use when not quoting myself in a family magazine. "I just got all my points back, after two bloody years without a ticket!"

But—miracle of miracles—the cop did not pursue. His brake lights eased off and he kept going, perhaps with larger fish to fry. Or larger cars to catch. Had we been thrown back, like an undersized trout?

If we had, it was all right with me.

Mike let me drive over the back hills of Wisconsin for more than an hour, and I liked the almost limitless grip of the Mini in switchbacks, and its easy, quick steering. As a great fan of the old Cooper S of the 1960s, I found the whole car a little more rubbery and detached in its steering and suspension feel than the mechanically taut original, but I guess that was to be expected. There is almost no car on earth as fun, direct, light and communicative as the original Mini, so it was a hard act to follow in a car that has airbags, crashworthiness and all the other modern baggage. Given those compromises, the new Mini is probably about as good as it can be.

Mike and I came back to the house in late afternoon, and he said, "Okay, now you and Barb have to take a drive by yourselves." He grinned at me conspiratorially over Barb's shoulder, like someone who had just dropped a few tabs of acid in the punch bowl and was proudly standing by to witness the inevitable transformations of personality that would soon take place.

Off we went for a short drive into the country, while Mike waited on our porch swing with a beer and our three confused dogs, who appeared to be wondering if we'd traded our home and dogs for a new Mini.

Barb had fun behind the wheel, but when we pulled into the driveway, she looked around at the rather flashy art-deco dash and door panels and said, "I really like driving this car, but I don't know if I could stand to look at this interior every day. It's too contrived. I like the simplicity of the old Mini better."

Which pretty much summed up my own thoughts. The car looked right on the outside, but they'd missed the uncluttered spirit of the original within. Too trendy and Euro, without enough British reticence.

Still, I thought, *if you really like driving a car, you can always look out the window....*

"Off we went for a short drive into the country, while Mike waited on our porch swing with a beer and our three confused dogs, who appeared to be wondering if we'd traded our home and dogs for a new Mini."

Also, there may have been an intentional message in that interior design. It said, "We know this car is supposed to be British, but you can tell by looking around you that modern Germans have been involved, with all the obsessive attention to detail that implies." Or, put more simply, "These people own micrometers!"

Mike, having accomplished his missionary visit, said goodbye and headed for Illinois to visit his mother for a few days, then cruised back home to Colorado.

A few weeks later, I got a call from Tom Harrer, an old racing buddy who used to drive a TR-4 and an S2000 in the SCCA. He told me he and his wife Anne were picking up their new green Mini in Milwaukee and would be coming through the Madison area. So we invited them to dinner, and Barb and I got to take a drive in their standard, non-supercharged Mini.

Nice car, and in some ways I liked it better than the S model, just because you have to work it a little harder to go fast. Simpler styling, too, more like the old Mini.

So we'd finally had a drive in both iterations of the new Mini, thanks to generous visitors, but this was not the first time a friend had attempted to spread the gospel.

Only a few months earlier, our friend Richie Mayer, had called us from Sedona, Arizona. Richie is a songwriter and music producer who restores old Alfas, vintage races a Porsche 914-6, rides a Moto Guzzi and has a small collection of vintage electric and acoustic guitars. Another perfect being, in other words, disguised as a citizen of Earth.

We mean no harm, but are awaiting instructions.

Anyway, Richie called and said, "Guess what? I've just won a chance to buy the first Mini Cooper S at the dealership in Phoenix."

"You won a chance to buy a car?"

"Yes. The dealership had a poem-writing contest about the new Mini, so I wrote a song about it and made a CD. The song is called, "I Kiss My Mini." I won the contest, so now I have a yellow Cooper S on the way."

Nothing like having a full recording studio in your house to overwhelm the best poetic efforts of some fourth grader with a crayon and a school penmanship tablet with Frodo on the cover.

A few weeks later, Richie called back to say he'd taken delivery of the Mini. "This thing really is fun," he said. "You've got to drive it."

I finally drove Richie's car last month. While hauling our dirt bikes to Baja, my buddy Pat Donnelly and I stopped at Sedona and went driving all over the beautiful Red Rocks area in the yellow Cooper S. Richie likes the car so much, it seems to have temporarily nullified his usual passion for buying and restoring hopelessly shot old Alfas.

"I'm looking at an old Alfa GTV," he told me, "but I don't know why I would drive it instead of the Mini. The Mini has so many things going for it; it's neat-looking, fun to drive and you can go anywhere without having to work on it. And it's new!"

I frowned hard and tried to grasp the possibility that those four attributes could all exist simultaneously in the same car.

Then, by way of anecdotal overkill, I ran into my friend Bill Neale—automotive artist, motorcycle aficionado, Texas gentleman and Cobra driver—at the Amelia Island Concours this year. He told me he'd bought a Cooper S and liked it so much he'd taken it on the Texas Hill Country Rally, instead of driving his Cobra.

Yesterday, I got a letter from my old pal Doug Harper, who is a sociology professor at Duquesne University in Pittsburgh. Another guitar player/car buff of impeccable credentials and subtle judgment, even if he has just only started riding motorcycles due to some delayed learning problem.

Doug informs me that, after much sleeplessness and soul-searching, he's trading his beloved Miata in on a new Cooper S.

So the pressure mounts.

I haven't quite decided yet if the new Mini is exactly the car for me and my countrified needs in the people-and-stuff-hauling department, but it's awfully nice to see people who love cars—and understand the mechanical essence of things—finding satisfaction and a spirit of affordable fun in something new.

It seems to me this happens only every five or 10 years. The Mazda Miata had that capacity to reawaken car enthusiasm among the faithful, and so did the Porsche Boxster in recent times. And now the Mini, which, I believe, passes the single most stringent test of good design: When you spot one on the highway, you are helpless not to point it out to others.

Your right arm levitates of its own volition and points at the passing car like a magnetized compass needle, and your voice automatically says, "Look, there goes a Mini!"

Forty-four years after their introduction, the old ones still do this too. It never fails.

September 2003

The Ferrari Sharknose

Not long ago, I was sitting at my word processor staring into space, as usual, and awaiting divine inspiration for my next adjective, when our three well-trained dogs began a deafening cacophony of vicious barks and mournful howls, a sound that must be heard to be believed. Barb and I call it "Baskervilling."

When the dogs bark like this, it usually means one of three things: 1) Armed invaders in black ski masks are entering our home; 2) There's a chipmunk sitting on a rock out near the sycamore tree; or 3) A delivery truck is coming up the driveway.

It was the FedEx truck this time, and the driver handed me a package from the California R&T office as all three dogs sniffed his right front tire. I hefted the padded envelope and squeezed it around the edges. "Feels like a book," I said, using the sort of keen deductive reasoning that has allowed me to differentiate one object from another (ball, chair, etc.) since I was a small child.

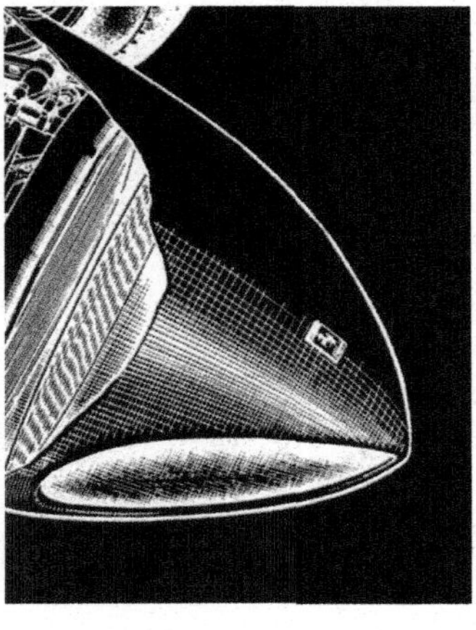

I took the package inside the house, sliced open the wrong end so all the gray sawdust fell out, and into my waiting hands slid a hardcover book with a beautiful picture of a red racing car on the cover.

Ferrari 156 Sharknose was the book's title, written by Ed McDonough, with a foreword by Phil Hill. There was a note from R&T Associate Editor Mike Monticello, asking if I would review the book.

I knew the cover photograph well. It was a famous picture by Julius Weitmann of Phil Hill at the wheel of a vividly red 156 Ferrari Formula 1 car, cranking the front tires hard left through Station Corner at the 1961 Monaco Grand Prix. Brilliant sunlight reflects off the nose of the car with its white number 38 and yellow prancing horse emblem above that menacing set of flared nostrils.

This identical photograph, incidentally, graces the cover of a Norton-Batsford paperback I bought as a freshman in high school in 1962 (cover price $1.50) called *The Racing Car,* by C. Clutton, C. Posthumus and D. Jenkinson. This book still resides on my office shelves, right behind a small glass-covered model of this very same Ferrari GP car.

And nearby is a framed front cover of the January 1962 issue of R&T that celebrates Phil Hill's 1961 World Championship with a photo of Phil driving the Sharknose car. I even ambushed Phil into signing it for me during one of our comparison test adventures.

In other words, Mike Monticello sent this book to the right guy. The Ferrari 156 Sharknose is simply my favorite Grand Prix car of all time.

Now, I realize that dedicated F1 buffs with a highly developed sense of history and aesthetics will probably roll their eyes or slap their foreheads (or both, simultaneously, if they are well coordinated) at that choice. How could anyone prefer this very handsome but simple and slightly tankish 1.5-liter car with its flared nostrils over, say, the magnificent Auto Union, Alfa or Mercedes GP cars of the 1930s? Or the lovely Vanwalls or Maserati 250Fs at the sunset of the front-engine era? Or any of a dozen higher-tech, winged, downforced, shift-paddled and brilliant designs since?

Well, because the Ferrari 156 Sharknose is Where I Came In.

The car is emblematic of my discovery of Formula 1 racing at the age of 13, and I'm not exaggerating when I say that the sight of that red car on the cover of McDonough's book as it slid out of the envelope and into my hands actually caused the hair to bristle on the back of my neck.

As good pictures of this car always do. It's an image that encapsulates all of the excitement, drama and deadly seriousness of Grand Prix racing at the very moment I discovered the sport through the pages of this magazine, and those of another excellent and now-defunct periodical called *Sports Car Graphic*.

"I read, oblivious to everything around me, with almost the same measured-breathing attentiveness as that 13-year-old who picked up his first R&T 42 years ago."

No Speed Channel in those days. In 1961, F1 got about as much television coverage in the U.S. as the Siberian Yak-Roping Championship, with the brilliant exception of a *CBS Reports* roundup of the 1961 season, tersely narrated by racing driver and enthusiast Walter Cronkite, bless his heart. Otherwise, there were only magazines, with wonderful race reports by Henry Manney III or Bernard Cahier.

Henry Manney III

And even those magazines were hard to find if you lived in a small town in the Midwest.

Luckily for me, my hometown contained at least one older College Guy who was a sports-car buff (Chris Jepson, I believe) who asked Lawrence's Drug Store to carry both *Road & Track and Sports Car Graphic*.

Druggist Ken Lawrence, a nice man of much natural curiosity, decided to go way out on a limb and order two copies of these exotic magazines each month. This is where I—already a budding Indy-car and midget racing fan—stumbled upon them.

I walked slowly up the hill to our house in September 1961 carrying copies of both magazines—for the total cost of $1.00, which I had just earned by mowing Mrs. Emmons' entire lawn—and looked carefully at the pictures and photo captions, which I found absolutely electrifying.

Here was real road racing, with slim, technically sophisticated cars slithering through the streets of Monaco or sweeping over the forbiddingly fast rural roads of Spa, past stone bridges, old farmhouses and pine trees. The danger of the sport was palpable in those brooding, black-and-white photos. There was almost a wartime aspect about them; you could see in the faces of the drivers they knew history was being made, at great, but unspoken, risk to themselves.

But there was also a carefree glamorous side, and the circuits themselves were lined with festive advertising banners for European products at whose uses I could only guess. What was Punt E Mes? Or Campari? I knew Dunlop made tires, but what about Ferodo? Could a person rub Ferodo into his hair, or was it a drink?

"Innes! Stirling! Come and join us at the bar! We are having an ice-cold Ferodo with Gendebien and Trips!"

I didn't learn until later they were brake pads. All quite exotic.

And most intriguing of all, three of the best drivers—Hill, Ginther and Gurney—were said to be Americans. And two of them—Hill and Ginther—were high school buddies from Santa Monica, California, and now they were stars of the all-conquering Ferrari team!

How could this happen? It was like discovering that Einstein and Freud used to skip class and shoot pool together, or that Grant and Sherman were in the same Cub Scout troop. What were the odds?

Exciting stuff, in any case. I read these magazines from cover to cover, many times, with that aching desire to learn and absorb information that only the young possess when some new vision of life sets their souls on fire. The hook was set, and at the very tip of the barb

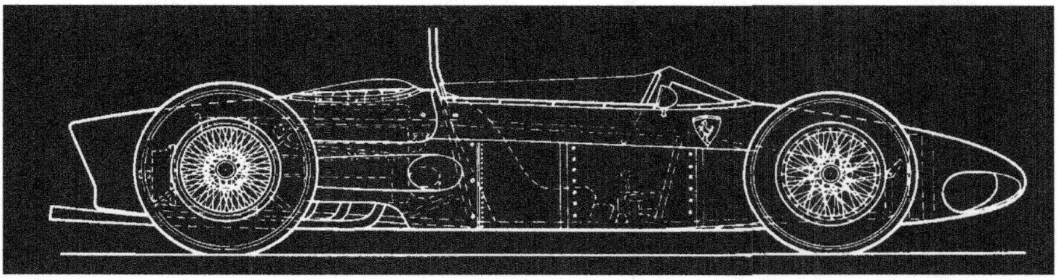

was that beautiful Sharknose Ferrari, being driven to victory by our first American World Champion.

So reading McDonough's book was no problem for me. As I mentioned in the review, it's well written, well researched and filled with fascinating interviews and anecdotes by nearly everyone who had any role in the design or driving of the 156 Ferraris. Going from cover to cover took about as much disciplined effort as falling down a flight of stairs, and I read, oblivious to everything around me, with almost the same measured-breathing attentiveness as that 13-year-old who picked up his first R&T 42 years ago.

Reading the book, of course, sent me into a sudden paroxysm of nostalgia for the 1961 GP season, and I was forced, late at night, to dig out my old racing videos that cover the early '60s F1 scene. Among these was a Speedvision *Phil Hill Retrospective*, and an old BP collection entitled *Moss the Master*, which has superb footage of the 1961 Monaco Grand Prix in which Ginther and Hill chase Moss, driving Rob Walker's Lotus 18, through the city streets in one of the great races of all time. It also covers the 1961 German Grand Prix at the old Nürburgring.

In the background, and in the pits, one can see images of our old and dear friend, the late Innes Ireland, looking like the madcap, slightly unpredictable character he was, and some new guy named Clark driving a Lotus in what looks like a leftover football helmet with no visor. And, of course, there's much footage of another racing hero of my youth, the aristocratic Wolfgang Graf von Trips, who lost his life at Monza while chasing the World Championship in another Sharknose Ferrari 156.

I watched these films and thought that if you had to be introduced to this sport at some time in the 20th century, you could do a lot worse than these drivers and those cars. I could hardly have asked for a better moment.

And the Ferrari 156—with no apologies—is still my favorite GP car. The earlier front-engine cars, though admittedly beautiful, will always look a little old-fashioned to me, while newer GP cars look...well, newer but not necessarily better. The Sharknose remains ageless to this kid, an agreeable spot of excellence, fixed in time.

October 2003

Ends of the Road

If there's anything I hate to smell while driving—other than the heady aroma of a melting TR-3 wiring loom, which is impressed on my memory forever—it's the scent of raw gasoline.

I've never actually had a car catch fire, but I am a great respecter of gasoline fumes, having singed my eyebrows into two abbreviated smudges of soot while trying to ignite a highly experimental rocket engine made of two Contadina tomato paste cans (air and fuel) joined by a piece of soldered copper tube when I was about 10.

"What happened to you?" my mother asked at the dinner table.

"Nothing," I said, typically.

And "nothing" is good, when it comes to gasoline fires. That's why I was a bit alarmed when I stopped by my friend Chris Beebe's shop, Foreign Car Specialists, the other day in my aged Buick Park Avenue and smelled fumes while parking.

I got down on one knee and found a spreading pool of gas beneath the car, just behind the left passenger door, where a fuel line had rusted out. This is what happens when you don't pay full price for a new Buick, but wait 25 years so you can get it slightly discounted, for $1800.

Chris saw what was going on and ran out with a bucket of water to dilute the fuel before it ate through his new blacktop and asked if I would park on the gravel side of the lot until one of his mechanics could take a look at it. He lent me his blue Miata for the day, and I was on my way.

Foreign Car Specialists, incidentally, doesn't normally work on Buicks, but I was able to slip it past the service department by speaking Italian with a Swiss accent. It also helped that I used to work there.

Chris called later to say the car needed an entire "cluster" of feed and return lines that supply the pressurized fuel-injection system. The fuel tank would have to be dropped to get at the connections. He said they could do the whole thing for about $300 and change, which sounded quite reasonable, considering the many parts and the amount of labor involved.

He stopped by our house later and traded the repaired Buick for his Miata. "Have you noticed," Chris asked, "that the brake pedal on your car just keeps going down if you hold it on hard at a stoplight?"

"No," I said, "But the pedal has seemed a bit spongy lately...."

The next evening Barb and I tried to drive into Madison and the brake pressure warning light came on. We returned to the house and I found a trail of brake fluid following the car into the garage. Slow leak from a ruptured—and rusted—brake line, it turned out. I topped up the fluid and Chris, who lives just across the creek, took it back to the shop in the morning.

"You might want to duct-tape a large pillow onto the front bumper," I suggested, "in case the brakes go out entirely and you have to rear-end another car to slow down."

"I'll take it easy," he said. Like most mechanically inclined people, Chris loves the challenge of nursing a wounded car through traffic without triggering a major explosion.

He called later to say all the steel brake lines in the car were pretty rusty and should be replaced, along with the flexible brake hoses. All this would cost about $300 plus change, which sounded quite reasonable to me, so I told him to go ahead.

You will notice that, as a former professional mechanic, I could probably have replaced these fuel and brake lines myself, but I decided at about the age of 50 there were certain jobs I just wouldn't do any more without a real overhead garage hoist. There's a huge difference between walking around upright beneath a car and rolling around on a creeper with a hot exhaust system grazing your nose. Especially if you're trying to bench-press a gas tank off your chest.

". . . the countryside is full of trusty old family sedans that gave up the ghost . . . and have been put—literally—out to pasture."

Anyway, the car got fixed without my expert help or advice and Chris dropped it off the next day. Now everything works fine.

For the time being.

But the clock is ticking. While driving the Buick yesterday, I noted that it now has 153,000 miles on the odometer, and reflected that things have recently been going wrong with ever-increasing regularity. A few months back, the crank sensor went out, and only weeks before that, most of the EGR system needed replacing, to the combined tune of about $1000. Then there were the new brake rotors (aftermarket)...which immediately warped, so the brake pedal now feels as if you're playing the kick-drum on a good set of Ludwigs.

A few small things have gone wrong too. I recently slammed the glovebox a little too hard and snapped the latch assembly off, breaking an inner plastic mount. I considered keeping the lid shut with duct tape, but decided that would make me feel too much like the victim of a Jeff Foxworthy redneck joke, so I found it could be wedged shut with the tasteful insertion of a chunk of blue rubber foam under the lip of the dash. Another elegant fix.

In other words, the old girl is nickel-and-thousand-dollaring me to death, and a quick perusal of my checkbook tells me I could just as easily be making payments on a new BMW 330i as fixing the elderly Buick.

So I am keeping my eyes open for another car, but I continue to drive the Park Avenue in the meantime, waiting for that final straw—the one expensive repair that causes me to say, "No more." It has to be coming.

Every car, unless it's a timeless classic with a number plate that's worth more than the car itself, eventually reaches the end of the road. It's a sad thing to think about, but there is always a moment when the key is turned in the ignition for the last time.

"Every car, unless it's a timeless classic with a number plate that's worth more than the car itself, eventually reaches the end of the road."

This is something I think about often, because we live out in the country where yard space is not at a premium and zoning is pleasantly lax, so local farmers often just park their old dead cars beside the barn. I'm not sure whether they're kept for spare parts or out of pure nostalgia, but the countryside is full of trusty old family sedans that gave up the ghost and have been put—literally—out to pasture.

You see a Ford LTD, a Chrysler K-car or a Chevy Citation (or all three) sitting next to a silo with that slight tilt toward a low tire and can't help wondering what particular malfunction caused it to finally be abandoned.

In most cases, I suspect something large and expensive went wrong—transmission, head gasket, etc. Or maybe it was a lot of little problems that made the car seem too unreliable for a family vacation. Some of our neighbors have given their older cars to their kids, who nurse them through college until they can afford something better. And then the car returns home, like a lost dog, to bask in the barnyard sun.

Whatever the reason, I like seeing these spent old cars parked on farms. It seems like a better end for them than a trip to the metal shredder or the stamp press. Conservationally sound as it might be, you hate to see these faithful runners flattened and stacked, six inches thick, on a flatcar bound for the steel mill.

I don't know what I'll do with the Buick when it stops running. If I had any sense, I'd quickly sell it (with the built-in "dirt-cheapness defense" against future complaints) or trade it in while it's still working fine. I suppose we'll eventually trade it in, just to get it out of sight and out of mind. Better if we don't know what happens to it after that.

Feels a bit cold-blooded, though.

When we were driving home from dinner last evening and discussing a replacement car, Barb looked around at the inside of the Buick and said, "Look at all this wonderful stuff that still works just fine. This car has beautiful leather seats, a nice sound system, perfect air conditioning, lights, wipers, heat...and all of those things will probably be lost forever, just because one small part finally fails."

I let that thought sink in as we drove through the night, and suddenly realized it's a concept to which I've become more sympathetic with age.

This afternoon, I drove the Buick to the doctor's office for my annual physical and found myself wondering whether cars—as extensions of our own brains and imaginations—are not actually a part of Nature, rather than mere collections of mechanical parts.

I resolved that after my appointment I'd stop at some new-car dealerships on the way home, just to look around. Sometimes it's better if we aren't exactly what we drive.

November 2003

Close Encounters of the Corvette Kind

A few weeks ago, my friend Jeff Zarth invited me to a big Corvette gathering in Madison, Wisconsin, called "Corvettes on the Isthmus." Jeff is a member of the National Corvette Restorers Society, and has a beautiful 1966 Laguna Blue Sting Ray coupe he exhibits.

I naturally accepted Jeff's invitation because 1960s' Sting Rays like his are longtime favorites of mine, and I also have a weakness for various versions of the chrome-bumper "Coke-bottle" cars from 1969–1972, as well as the modern C4 and C5 convertibles, which always seem like a tremendous bang-for-the-buck value to me in handling and horsepower.

So Jeff and I walked around and looked at cars all morning, then had lunch at an outdoor hamburger stand. "Any new or interesting cars in your stable?" Jeff asked.

"No," I admitted. "I finally got the Jag E-Type back on the road, but I've been looking for a sensible, commuter-type car to run around in. Something that gets really good mileage, doesn't cost too much and doesn't need to be worked on."

Without blinking an eye, Jeff said, "Why don't you buy a used Corvette?"

I laughed out loud and said, "Well, I like Corvettes, but they aren't really the sort of practical car I need to offset the Jaguar, if you know what I mean. Plus, I don't want to spend that kind of money."

Jeff shrugged. "Well, I know a guy who's selling a really clean 1987 convertible with the manual 4-speed +3 overdrive transmission. It's dark red with a tan interior and top, and it has just 27,000 miles on the odometer. He's only asking $13,000 for it. I'd buy it myself, if I had room in the garage."

My eyes narrowed, my neck shortened slightly and my jaw went all rigid. Same reaction you get out of a person who suddenly hears a bullet whistle overhead."Do you have his number?" I asked.

Notice how easily I was distracted here from my original plan. It happens over and over again.

Me: "Do you have any sensible brown walking shoes?"

Shoe salesman: "How about a nice rocket-propelled backpack instead?"

Me: "Let me see it."

"He insisted I take the Corvette out for as long a drive as I wanted."

So of course I got the number and called the Corvette owner. Had to. I'd been looking at perfectly nice little commuter cars in the $14,000 to $19,000 range, and here was a C4 Corvette—a car with 245 bhp, giant grippy tires and superb suspension, designed, in its time to wallop Ferraris and Porsches at the racetrack, all for $13,000.

The owner, a friendly and pleasant guy named Dave, had a lovely home on the lake, and he insisted I take the Corvette out for as long a drive as I wanted. So on a warm summer afternoon, I found myself motoring along the shores of Lake Monona and then out onto the highway to check out the car's highway cruising manners.

The Corvette had comfortable seats, a great sound system, lots of easy horsepower on tap and a nice, relaxed gait on the highway with 4th overdrive engaged. Ride was moderately good, and on exit ramps it had more grip than some of us have neck muscle. Wind flow at speed was quite serene with the top down. A big, friendly, fast, easy-to-drive car with nicely balanced steering.

I pulled over on a side street and got out to contemplate the car—something I always do on a private test drive.

The Corvette was such a nice car for $13,000 that I found myself chuckling at the sheer illogical extravagance of most of the other sports cars I'd owned. Was it really possible I'd spent $1000 more than this, 10 years ago, for a totally worn-out Porsche 356B that needed absolutely everything replaced or rebuilt?

And I won't even get into the economics of redoing my E-Type. Let's just say I'm into it about three Corvettes' worth.

"The Corvette was such a nice car for $13,000 that I found myself chuckling at the sheer illogical extravagance of most of the other sports cars I'd owned."

Yet here was a fast red sports car with no evident problems, apparently ready to drive to the West Coast and back, top down, at a moment's notice. No restoration needed. Granted, the old Jag and the Porsche were rarer and had higher entertainment value for bystanders, but the Corvette was a car for entertaining yourself—and one close friend. For which there is often much to be said.

When I got back to Dave's house, my instinct was to whip out my checkbook and buy the car, but instead I cautiously told him I had to think about it and talk to my wife, Barbara.

"Let me know what you decide," he said.

When I got home, I said to Barb, "Well, I'm glad you didn't go with me to look at this Corvette."

"Why?"

"Because you would have wanted to buy it."

"So why didn't you buy it?"

"Because we already have a 2-seater sports car. And when winter comes, the Corvette won't be any good in the snow and we'll have two cars to store for the winter. And because a Corvette convertible has even less luggage space than our old Miata. You almost can't carry anything."

"It sure would be fun to take on a summer vacation..." Barb said in a small, lost voice, "or to drive up to Door County on the weekends when we go sailing...."

Unspoken in that expression of longing was the fact that our old Jag didn't have air conditioning and sometimes made "funny noises" at random times. Last week when we drove up to Lake Michigan, for instance, the speedometer suddenly started going *woo woo woo*, exactly like Curley of the Three Stooges, and shortly thereafter the needle started to dither around like a compass in an iron mine.

The Corvette, being newer, presumably would not do this. For Barb, it represented a chance to travel without suspense. And the sports cars I normally restore and own are nothing, if not suspenseful.

But the next day, in the clear air of morning, I decided not to buy the Corvette. Couldn't really afford it, and it was the wrong car at the wrong time. Not what we needed. Barb took the news stoically, but I sensed disappointment.

Strangely, this was the second time in our lives we'd come perilously close to buying a Corvette.

The other time was in 1973, not long after we were married. Barb and I decided to go car shopping to replace our high-mileage Volkswagen Beetle, and one weekend made the rounds of all the dealerships in Madison. We looked at all the usual econoboxes of the time—Pintos, VW Beetles, etc.—and eventually ended up at a Chevy dealership because I wanted to see the new 100-bhp Vega GT.

We took one out for a drive, and found it to be a tinny, cheap-feeling car with a reasonably zesty engine as its principal virtue. We thanked the salesman and started to leave, when he said, "Say, you folks seem to be drawn toward sporty cars. You interested in a Corvette?"

We turned and looked at him. "They're a little out of our price range," I said.

"Maybe not," the salesman replied. "With this fuel crisis on right now, nobody will even look at a big-block Corvette, but we've got a beautiful low-mileage 1970 454 Stingray coupe. Green with a tan interior, 4-speed, side pipes.... We could probably let you have it for around $3800. We just got it in on trade, so it's back in the clean-up bay right now."

Barb and I went back and looked at it. The car was immaculate. We took it out for a test drive and it was so fast and exciting to drive—especially after the cars we'd just driven—that we actually broke out laughing. It was hard to imagine that the Corvette and our rusting Beetle were both "cars," in the same way it's odd to think that Chihuahuas and Great Danes are both "dogs."

We lunged from one street to another, blasting ourselves from point to point in a bellow of back-slamming acceleration and glory until we came back to the Chevy dealership.

Barb said, "We could buy the Corvette and drive up north to my parents' cottage this weekend. My dad would love this car."

"Your dad would think I'd lost my mind and his daughter had married an immature idiot with no financial sense."

"No, he wouldn't..." she said, a little sadly.

After a sleepless night, I decided not to buy the Corvette. There was a fuel crisis on, I was spending most of my meager income racing an H-Production Bugeye Sprite, we were still paying off college loans and we needed a "real" car, a winter car for groceries and laundry and commuting. I told Barb I thought it would be better if I rebuilt the engine in the VW, painted the rusty fenders and kept it going for a few more years.

Which I did. Seven more years, to be exact.

But when I told Barb we shouldn't buy that lovely green 454 Corvette, I thought I noticed just a little bit of light go out of her eyes, as if I'd failed some private test she'd devised for me, broken some unspoken pact that our lives would always be unpredictable and fun.

And I saw that exact same fade to gray last week when I decided to pass on the dark red 1987 convertible, that same quiet flight of joy and anticipation from her expression.

Maybe I'd better buy the next good Corvette that comes along. This one feels like strike two.

December 2003

My Old Pal, Clark Gable

My older sister, Barbara Card, who is a nurse from Irvine, California, suddenly became a patient recently. While leaving work at the hospital, she crossed the lawn toward her car and tripped on a sprinkler head. Unfortunately, she fell onto the sidewalk and broke her right arm in many places.

Before you jump to any conclusions that my sister is a clumsy person, I should say that, when we were in high school, she was easily the more athletic of the two of us, a genuine A-team cheerleader, while I was a merely adequate football player. And I use the word "adequate" in its most flattering sense.

Still, you have to wonder about a person who trips on a lawn sprinkler. You'll notice *I've* never done that. At least, not on a running one.

I won't rub it in any further, though, just because my sister was so mean to me when we were kids and used to put her hands and feet over on my half of the car seat when we were on road trips. Also, I still think it might have been her cigarette rather than mine that set the Milne's field on fire.

But that's all water over the dam—and out of the fire hose—and these days, of course, I love my sister dearly, so I invited her to spend part of her convalescence visiting my wife Barbara (a.k.a. Barbara Egan II) and me in Wisconsin. I promised we would wait on her hand and foot, cook meals, drive her around to stores and shopping centers and all the other things she can't do with her arm in a cast.

So she arrived last week at the Madison airport, carrying an X-ray of all the nuts and bolts and plates in her arm to explain the metal detector alarms going off everywhere. Nevertheless, airport security still put her through the mill, going through her bags and making her take her shoes off.

A lot of people think it's young male religious lunatics from the Middle East who are causing all the problems, but apparently it's not. It's blonde, middle-aged mothers of three from Irvine, California, with nursing degrees and broken arms. One look, and you can see they're nothing but trouble.

"I was amazed that someone had been able to build a car that genuinely resembled the very thing for which it was named. This didn't happen very often; our neighbor's old Studebaker Dictator, for instance, didn't look anything like Mussolini."

Anyway, my sister arrived without detonating anything, and we had a wonderful visit, staying up late and talking about the old days, books we're reading, recent movies, classic movies we've seen on TCM, new cars we like and so on.

As we talked into the evening, it slowly dawned on me—as most things do—how much of my life-long car addiction can be traced to my sister's not-always-corrosive influence.

Barbara is four years older than I, you see, and at a time when I was still trying to put the green wooden roof slats on a Lincoln Log cabin or doing dangerous things with sunlight and a magnifying glass, my sister was reading movie magazines. Stacks of them.

She also had movie star scrapbooks, and filled them with photos and articles from newspapers and fan magazines. Prominently featured, as I recall, were Montgomery Clift and James Dean, and my sister—to her eternal credit—was a confirmed Elvis Presley fan during the great Pat Boone/Elvis schism that was sweeping teendom at the time. She had several large Elvis scrapbooks, and bought all his records.

So, at the tender age of 8, I got exposed to "Hound Dog," which I

thought was the greatest piece of music I'd ever heard (and it may still be), and I listened to my sister's record of it, literally hundreds of times, on the small record player in my room.

There is much to be said for having an alert 12-year-old sister when you are still in the gill stage of your cultural development. It's like sending out a scout from the wagon train to find out where the water is.

But there were other payoffs besides the music.

One day, well before Elvis came along, my sister showed me a picture in a movie magazine of Clark Gable standing next to a Jaguar XK-120.

"This car is called a *Jaguar,*" she said. "It's a sports car made in England."

I stared at the photo of that car and was absolutely smitten by it. I'd never seen a sports car of any kind, and this thing actually looked like a big cat—a svelte, crouching thing on its low wheel spats, ready to leap down the road.

I was amazed that someone had been able to build a car that genuinely resembled the very thing for which it was named. This didn't happen very often; our neighbor's old Studebaker Dictator, for instance, didn't look anything like Mussolini.

Anyway, the Jaguar embedded itself in my brain as the

best-looking car I'd ever seen, and I later spent the savings from my lavish 25-cent allowance on a red Dinky Toy model of the Jaguar XK-120. I could spend hours holding this model up to the light and admiring its lines, even during arithmetic class at school. Especially during arithmetic class.

Through my sister's James Dean fixation, of course, I was also exposed at a tender age to the Porsche 356 and the fatal 550 Spyder. Dean's death was a tragedy, but the intoxicating blend of danger and those exotic, beautiful Porsches was not lost on me. I spent a long time looking at the pictures and trying to read the stories without moving my lips. Hence, I learned about Porsches.

And to my sister and her movie magazines I owe one other debt. She was the first person I ever heard say the word "Ferrari" aloud.

In the 1950s, it was quite common for well-heeled Italian film stars and directors to compete in the Mille Miglia, and Barbara showed me a photo of the great director Roberto Rossellini, sitting in his Ferrari (if memory serves from this distance) on the starting ramp in Brescia, kissing Ingrid Bergman goodbye as he prepared to leave on 1000 miles of unimaginable adventure and danger on the winding roads of Italy.

And, of course, when the Marquis Alfonso de Portago was killed in the 1957 Mille Miglia, there were many before-and-after photographs of his Ferrari in Barb's movie magazines because Portago was romantically linked to some young movie star. In any case, the whole magic of the Mille Miglia and the very existence of cars called Ferrari were made known to me purely through the agency of those movie magazines and my sister's scrapbooks, years before I saw my first copy of *Road & Track*.

It's easy to dismiss the importance of movie stars in the sports-car and racing world as trivial and frivolous, next to the true gravity of, say, a Fangio or a Moss, but these celebrities certainly had a beneficial effect on my own outlook. How else would a third-grader in a small Wisconsin town in the '50s ever have been exposed to Jaguars, Porsches and Ferraris, and the people who drove them? Or the exotic places in which they were driven? It's hard to imagine another source.

But, in any case, it was the cars, rather than the stars, that mattered most. I admired the Jaguar 120, not because I wanted to be just like Clark Gable (fat chance), but because Clark Gable made the car known to me. From my point of view, he was just another guy who loved cars but happened to be famous. And, from what I understand, that might have been Gable's view of himself as well. He was merely a colorful part of the enthusiasm and good times that revolve around cars and make this crazy life more tolerable.

So—much as I hate to admit it—I probably owe my sister an incalculable debt in all of the life-long pleasure and activity I've had because she took just a few minutes of her time to dazzle my poor young brain (just then subdividing into two identical cells) with images of racetracks and exciting cars.

Later, when I became a fanatical car nut in my teens, I tried to pay her back by constructively using all the knowledge and insight I'd gained by reading car magazines almost continuously—especially during math class.

I told Barbara, when she went off to college and began serious dating, that she had only to tell me what kind of cars her boyfriends liked, and I would confirm whether or not she should consider a proposal of marriage.

She came home one weekend and told me she'd met a fellow drama student named Russ Card who had a brand-new 1966 dark green Mustang GT Fastback with a high-performance engine, a 4-speed transmission and a black interior.

Suppressing an involuntary cough, I slowly and deliberately held two thumbs in the upward position. "My professional advice is to marry the guy as soon as possible," I said.

She did.

And now, 37 years later, I have three nieces, Catherine, Christine and Anne, who are also magnetically—perhaps genetically—drawn to fun and interesting cars. Russ is no longer with us, but I think he would be proud.

And so would Clark Gable, for that matter. We think of him around here as almost a member of the family.

January 2004

Empty Garage Syndrome

My old friend and editor Allan Girdler has a great memory for literary quotes, and one of his favorites comes from the redoubtable Dr. Samuel Johnson, who, according to Allan, said, "Nothing is more terrifying than a Scotsman on the make."

One has only to look at the rags to riches career of, say, steel magnate Andrew Carnegie, to get the good doctor's drift. There is a flinty mercantile instinct at work here, combined with a Presbyterian work ethic, boundless energy and an unwavering desire to succeed one often finds in lands with scant resources but a tightly knit social structure. The whole picture is one of irrepressible force of will. Terrifying, indeed.

However, I recently thought of one thing that is even more unstoppable and terrifying than a Scotsman on the make, and that is a car buff who has money in the bank and an empty spot in the garage.

I ought to know, because I recently sold off a couple of motorcycles. One of them was a Harley-Davidson Electra-Glide Standard, a motorcycle whose market value is about the same as a new Honda Civic, and it was a big bike. A Harley with a fairing and saddlebags, tilted on its sidestand, takes up about as much floor space as, oh, say, a Lotus Elan, just for a handy example. In other words, the hole this bike once filled has begun to suggest all kinds of clever ways to fill it.

Yes, I have money in the bank for once (rather than the usual willingness to accept crushing debt) and empty garage space, both at the same time. Scotsmen, stand aside!

Naturally, the day I sold the Harley my phone rang before I'd even deposited the check, and it was an old racing friend from Chicago who was moving to smaller digs and decided to sell his largely disassembled

Jaguar XK120 project car—cheap. Real cheap.

"Okay, I'll drive down there tomorrow to look at it," I said.

That evening, I got out my large collection of Jaguar books and looked at every picture I have of XK120s, including the XKs Unlimited catalog of parts and prices, with exploded views of the car's mechanical innards. Useful for contemplation, and also to possibly strike fear in the hearts of sane people.

Didn't look too hard to work on. Especially after the E-Type. Some similar parts, but fewer of them.

Then I took the next step, which was to call my buddy John Jaeger in California, just to bounce the possibility of XK120 ownership off his car-fertile brain. John, even more than I, is quickly seduced by interesting cars, and, last time I looked, he had a TVR S2 Vixen, a Genie MK-5 sports racer, a Morgan Plus 4 drophead coupe, a Lotus Europa JPS Twin-Cam, a 1964 Ford Galaxie convertible and a 1965 Mercury Park Lane convertible parked somewhere around his property. In other words, he's crazy as a bedbug.

Still, he has good financial sense and a firm grip on market values—for a guy who needs a forklift to move most of his cars around—so I always call him with deals like this.

"Sounds too cheap," he said. "There must be something wrong with it. Still, you'd better go take a look."

"I'm going tomorrow," I said.

"Even if the car is a hopeless project," John said reflectively, "you've already enjoyed the best part of any Jaguar restoration, which is to get all excited and look at the pictures before you actually have to work on the car."

It's for words of wisdom like this that I call John.

The next day, I drove to Chicago and, as it turned out, the XK was a little too rough even for my masochistic tastes. It had a fair amount of structural rust, and I simply don't attempt restorations on rusty cars any more. I was hit across the side of the head with that two-by-four one too many times in my youth, and have grown easily spooked, like a horse that's been struck by lightning 10 or 12 times while standing at the same water trough.

"Maybe the best part of any restoration, really, is to get all excited and look at the pictures before you actually have to work on the car."

So the XK120 is out of the picture. Kind of a relief, really. I like 120s, but it wasn't right at the top of my Short List of Cars I'd Like to Own. As a garage mate to my recently restored E-Type Coupe, I picture something smaller—or more frail and archaic—to complement the elegant and fast Jaguar, something along the lines of an MG-TC or a Morgan Plus 4. Haven't ruled out a Series I or II Lotus Elan, either.

Nor a Velocette Thruxton. A friend of mine has one of these for sale right now, and the knowledge is eating into my restless sleep.

The right 1965 or 1966 Mustang Fastback GT is still on my short list too. But then so is another Norton Commando. Or a Brabham Formula B vintage racer. In other words, a lot of different things could end up in that empty garage space.

And of course John Jaeger keeps calling and directing my attention to cars for sale on eBay. In the past couple of weeks he's alerted me to several TCs, Mustangs and Morgans that have not yet been bid to their reserve prices. So far, however, none of these has quite grabbed me. There's always one thing wrong—color, price, location…something.

As if these temptations were not bad enough, yesterday I drove into town and spotted a pleasantly shot 1942 Dodge half-ton pickup for sale in a farmyard. A classic from the "Job-Rated" era, and one of my favorite designs. Not much rust, no interior, a set of Pinto (or something) bucket seats sitting loose in the car. This caused me to run right home and read my *Buyer's Guide to Dodge Pickup Trucks* and to flop on the sofa for the evening to reread John Jerome's always enjoyable book, *Truck*, in which he restores a 1950 Dodge pickup and waxes eloquent on the process.

I've restored a lot of old British sports cars; why not a pickup? A close examination of the old Dodge revealed it to be, basically, nothing but a taller MG-TC with useful hauling space, a flathead 6-cylinder engine and fewer whittled parts.

Tempting.

But then everything is tempting. To have money in the bank and a bit of garage space is to look at the world's cars from a lordly perspective. And therein is a bit of a problem.

Once you have passed judgment—and demurred—on the purchase of a Jag 120 with rust or a Dodge pickup with Pinto seats, you become a bit jealous of your freedom to look around, and you also become stingy with your funds. It's so much fun looking at old cars and motorcycles (not to mention airplanes and boats) that you hesitate to unleash your savings on a single vehicle and to find yourself, once again, as an optionless person with some headache-inducing project and no garage space.

When you are still in the hunt, the world is your oyster. Once you've spent the money and filled the space, the oyster can go bad quickly and the fun of the hunt is over. John is right—maybe the best part of any restoration, really, is to get all excited and look at the pictures before you actually have to work on the car.

So I am moving very slowly to fill that empty garage space. Becoming haughty almost, starting to think in the "royal We," like one of the kings of England: "We are not amused by your car. Please take it away. Bring another for me to look at."

Money and garage space are the catbird seat.

And, speaking of the catbird seat, Barb and I spent the day yesterday with our friends Tony and Kris Buechler, Tom

Severson and Karl Engelke at the annual Brodhead, Wisconsin, antique aviation fly-in, walking around the grass strip and looking at beautifully restored old Wacos, Ryans, Cubs, Taylorcrafts, Pietenpol Air Campers and the like.

I noticed that Tony, who recently sold his and Kris's Beech Baron twin and now has money in the bank and a big empty spot in his hangar, was looking at old airplanes with an especially bright light in his eyes, a certain energetic jauntiness in his walk.

I was just feeling the unquiet stirrings of possible airplane ownership again myself, when Karl (as if reading my mind) asked, "When are you going to get another Piper Cub and get back into flying?"

"If I got another airplane," I said, "it would probably be a 150-horse Citabria or Decathlon—an aerobatic airplane with real radios and some climb performance."

At that moment I turned and looked at a beautifully restored Beech 18 with its two Pratt & Whitney radials gleaming in the late afternoon sun. Written just below the pilot's side window, in lovely nose art script, were the words, *What was I thinking?*

Whether I end up with an airplane, a motorcycle, a pickup truck or a spindly British roadster with those funds that are burning a hole in my pocket, I am sincerely hoping to avoid—just for once in my life—a project that inspires those terrible words. This time, I'm making an effort to think long and spend reluctantly. As perhaps a Scotsman would, if he were on the make.

February 2004

Our Friend Gil

I think it's significant that when my friend Gil Nickel died of cancer at the age of 64 in California last fall, no fewer than four people from the Napa Valley area called to let me know he'd slipped away during the night.

Gil was a winemaker, racer of vintage cars, avid motorcyclist, pilot and yachtsman, and he had such a broad spectrum of friends and acquaintances, the phone grids must have lit up all over America and parts of Europe—the modern version of the church bell tolling in the medieval village.

Longtime R&T readers may recognize the Nickel name, because he was quite a friend to this magazine. We did a Salon article on Gil's 1953 Bentley R-Type Continental in 1990, and in 1995, Gil invited me to co-drive his bright yellow 1951 Ferrari 340 America Vignale Spider in the Mille Miglia. We did a story on that adventure, beautifully photographed by John Lamm and Richard Baron.

And, back in 1994, Gil invited me to co-drive his very fast and well-prepared Lotus 23B in the Chicago Historic Races at Road

America, and John Lamm and I did a story on that car too. So Gil and his cars have been in these pages a lot, and his fearless generosity gave this unworthy reporter some of the best moments I've ever had behind the wheel of a car. The odd thing is, I almost didn't meet him at all.

Way back in 1989, Editor Tom Bryant walked into my extremely well organized office at R&T, found me typing away behind some stacks of unanswered mail and asked, "How would you like to write a Salon on a Bentley R-Type Continental? It belongs to a guy named Gil Nickel, who owns the famous Far Niente Winery up in the Napa Valley. He's got a very nice car collection, and he's willing to let us drive the car, which was the fastest sedan in the world during the early '50s."

I let out a low whistle and gave Tom what may have been a slightly rueful look. I liked the idea of writing about a Continental R, but I was working on two other stories at the time. Also, I may have had a slight attitude problem.

An attitude problem about such an engaging assignment?

Yes.

It was the ghost of my egalitarian Midwestern upbringing that was raising its doubtful head. Naturally, I'd never met a Napa Valley winemaker, but I'd seen them on TV. What if this guy Nickel turned out to look like someone from *Falcon Crest* and walked around wearing an ascot and talking as though he had a pencil clenched in his back teeth? Could make for a long weekend.

Fortunately, John Lamm, who had met Nickel, assured me that Gil was "a great guy." He said, "You guys will hit it off immediately; you like all the same cars. Besides," John added, "he's a good old boy from Oklahoma."

"Ah," I said. Things were looking better.

So I took the assignment and,

like Peter Falk on a *Columbo* case, showed up at the big iron Far Niente vineyard gates with my notepad, but no trench coat.

Of course it took about 30 seconds for Gil to completely disarm me. He showed up just a few minutes late for our meeting, wearing cowboy boots, blue jeans and a slightly wrinkled checkered Western shirt, apologizing profusely for his lateness. "Gol' dang it!" he said in his enthusiastic Oklahoma accent, "ah been at Sears Point all day testing the Lotus 23B, and I'm just sweaty as an old pig. You'll have to forgive me."

"And it was that kind of party spirit and unblinking generosity for which Gil will probably be best remembered."

Now, if there's one thing I can forgive a person for, it's being overheated because he's been testing a Lotus 23B, my favorite sports racer of all time. Also, Gil's down-to-earth friendliness and upbeat energy were so infectious, I found myself laughing as I shook his hand. We've been friends ever since.

Gil gave me a tour of the winery and said he'd bought it in 1979 as a defunct vineyard that had closed its gates with Prohibition in 1919. He'd replanted the vine stock, restored all the old stone buildings and was about to build a set of wine caves (since completed) back into the hillside. The name Far Niente, he explained, was bestowed by the Italian family who founded the vineyard in 1885, from the phrase *Dolce far niente*, which means, "It is sweet to do nothing" in Italian.

"Italian farmers say this," Gil explained, "when they sit back and look at their fields after a hard day of work, and maybe sip a little wine."

On the shaded hillside above his home and office, Gil showed me a barn with his collection of 15 or 20 cars in it. This was not some suddenly acquired group of spotless classics, but a series of cars he'd owned and driven over the years—an Alfa Romeo Giulietta, Jaguar E-Type, Porsche Speedster, Jaguar XK-120, Ferrari Superfast, '32 Ford High-Boy roadster, and so on. He also had several BMW motorcycles and a couple of restored wooden boats.

Gil told me he bought his first new Alfa when he was just a young man, and the dealer told him to bring it back for an oil change and valve adjust when it had accumulated 2000 break-in miles. He picked the car up on a Friday afternoon and brought it back Monday morning for service, having driven around Oklahoma almost continuously, day and night, for the entire weekend. "I loved that car," he said.

How did an Alfa nut from Oklahoma come to be a Napa Valley winemaker?

Gil said he'd grown up on the family tree nursery in Oklahoma, which grew into the second-largest family-owned nursery in the U.S. He thought of becoming a minister, but decided instead to study mathematics (fortunately for the church, he said) and he became a missile scientist, plotting trajectories for the U.S. Navy Laboratory in Corona, California.

In Memory of...
Gil Nickel

"Wherever Gil happened to be standing, nothing was ever ordinary or dull for one minute."

He returned to Oklahoma to help run the nursery for a few years, then decided to learn winemaking, taking courses at UC Davis in northern California. He made his first experimental batches of wine in a basement apartment in San Francisco, and was soon looking for an old vineyard to restore.

"I used to fly back and forth from California to Oklahoma in my old Taylorcraft," he said, alerting me to yet another common interest.

That first evening in the Napa Valley, I went out to dinner with Gil and met his wife-to-be, Beth Yorman, another Oklahoman (by way of Wisconsin), and I was struck by the genteel, gracious charm of both of them. They shared a sort of courtly, Southern sense of hospitality in which no effort is spared to make a guest comfortable, and yet they both had a hilariously irreverent sense of wit and good fun. Gil could tell jokes all evening, and he also had a knack for composing funny, self-deprecating anecdotes that were perfectly emblematic of some larger meaning. His stories always stayed with you.

Barb and I gradually became good friends with Gil and Beth, and they introduced us to the Napa Valley Touring Society, a lively, hard-riding group of a dozen or more like-minded friends who took motorcycle tours every year. Together, we toured the Alps, Colorado, northern Mexico, the Ozarks, the Canadian Rockies and Oregon. Like the Nickels, the people in this group have become some of our best friends. So much for *Falcon Crest*.

Besides motorcycles, Gil and Beth liked boats and had a restored wooden Belle Isle Bearcat cigarette boat, as well as a huge classic steel motor yacht, *Lycon* (since renamed *Far Niente*), built in Holland before World War II. One year, Barb and I sailed with the Nickels under the Golden Gate Bridge and down to the Monterey Historic Races in this little ocean liner. They held a big Saturday night party on the boat for Gil's fellow racers.

And it was that kind of party spirit and unblinking generosity for which Gil will probably be best remembered—along with his winning wines, cars and competitive spirit—among his fellow vintage racers.

Every year at Road America, the Far Niente transporter showed up, driven by Gil's racing mechanic and sea captain, Chris Chesborough. They would unload a couple of Gil's cars (Lotus Elan 26 R, Lotus 23B, Lotus 27 Formula Jr., or Surtees Formula 2 car), and Gil would go out and usually set fastest time and win his race.

Toward evening, the Nickels would throw a party under the awning next to the race trailer, lavishly supplied with food and endless bottles of their own wine. This got to be such a tradition, I think most of us developed a Pavlovian response to the closing of the track at dusk and found ourselves wandering helplessly in the direction of the Far Niente transporter.

It was a nice place to be, and could be located by the sound of laughter. Wherever Gil happened to be standing, nothing was ever ordinary or dull for one minute. He was just a guy who brought his own fireworks, wherever he went.

And I gradually came to understand that, by reluctantly taking Tom's Bentley assignment years ago, I'd unwittingly doubled the amount of pleasure and fun this life has provided through racing, motorcycles, good friends and the surprisingly small universe of people who really love sports cars. It's quite a fragile and provisional structure, held together mostly by imagination and energy.

And now that Gil is gone, it feels like the lights have gone out in part of our house.

March 2004

Driving the Jaguar

This past summer, I went to my favorite vintage racing event of the year, the Brian Redman International Challenge at Road America, and spent some time hanging around the *Road & Track* tent, talking to quite a few readers of this magazine.

Some who read this column said they were curious about the fate of my 1967 Jaguar E-Type coupe.

"You've done a bunch of columns about restoring it," was the comment, "but haven't mentioned whether it's done or what it's like to drive now."

I frankly admitted I'd done so many columns about the various stages of this three-year restoration and its vicissitudes and trials, that I was hesitant to mention it even one more time, for fear of appearing to be squeezing yet another drop of blood out of the same old stone.

"If I say one more thing about the Jag," I told one guy, "I'm afraid that irate villagers with torches and pitchforks will appear in my driveway and burn the garage down, some bearing signs saying, 'SHUT UP ABOUT THAT STUPID JAGUAR,' or 'ENOUGH, ALREADY!' "

"Quite the contrary," a couple of hardy readers told me. "The E-Type is my favorite car, and I'd like to hear how it turned out."

Well, you don't have to ask me twice. Or, sometimes, even once. So, with this groundswell of plaintive supplication by nearly three readers—and that's a conservative estimate—I decided to bring things up to date.

In fact, the E-Type is done and on the road.

I got it fully operational more than a year ago, in the autumn of 2002, just about two weeks before the first snowfall (and salted roads) forced me to park it in my workshop for the winter. Those first shakedown runs were surprisingly trouble-free. I had a stuck needle and seat in the center SU carburetor that flooded the engine compartment with gasoline and forced me to do a roadside repair by blowing a piece of grit out of the passageway, and I had a bad front suspension rattle that I eventually traced to a missing washer behind a lower rubber A-arm bushing. That, and every single cooling-system hose seeped antifreeze until I cranked down extra-hard on all the hose clamps.

Once these minor annoyances were cleared up, the car ran perfectly and has been essentially trouble-free through the entire past

"'If I say one more thing about the Jag,' I told one guy, 'I'm afraid that irate villagers with torches and pitchforks will appear in my driveway and burn the garage down.'"

summer of driving. Barb and I have taken three or four long weekend trips in the car, driving to, among other places, the races at Road America and up to Wisconsin's Door County peninsula to Ellison Bay, where we now keep our new/used Com-Pac 27 sailboat in a slip. We use the Jag to drive into the city on nice evenings, and have taken many rides in the nearby hilly rural countryside, just for the fun of driving around. We've put about 4000 miles on it, so far.

We'd also planned a long, late-summer trip to New England, but that fell through when I was diagnosed with hepatitis C last May and had to stick around home to take six months of innumerable injections that make me feel as if I've either got the flu or been hit by a large truck. Or perhaps hit by a large truck loaded with 55-gal. drums of influenza virus. Which then backs up and runs over me again, for good measure. In any case, long trips have been out of the question, and, as of this writing, I have four weeks of this draconian, medieval drug therapy left to go.

When it's over, drinks will be served.

But enough about my defective self; back to the Jaguar.

So now that this baby is done, how is it to drive?

Well, I love it. A few observations:

1) The E-Type's mechanical sophistication is a big step up from the many other British sports cars I've restored, and everything about it is just a little nicer than it would have to be. The throttle linkage, for instance, rolls on real bearings instead of bolts shoved through holes. You can feel this when you drive. Sir William Lyons cheaped out on a few components (the instruments and switches are only fair) to keep the price competitive, but not many. The car is a mass of high-quality engineering solutions that give it a wonderfully satisfying mechanical feel.
2) Suspension compliance and steering feel are superb. Most Jags have always had these almost magical qualities of civilized ride and body-roll-free handling combined, and the E-Type has them in spades. No coal-cart British suspension here; just real road-holding through refined geometry and motion.
3) It's the first old British car I've owned (or old car, period) with real horsepower. When some yahoo with an elevated pickup truck starts tailgating you on the highway, you simply depress the accelerator and leave said rube gasping in a light tan cloud of unburned 1960s' hydrocarbons, wondering where you went. It has both a

high top speed and scintillating acceleration equal to nearly anything on the road.

4) The engine is unbelievably flexible. The old road test comments about leaving the E-Type in top gear as you drive through a village are true. That 4.2-liter twincam-6 pulls like a truck from idle on up, with never a sign of pinging or bucking, even breathing through those three big SU carburetors. It's simply the finest engine I've ever had in a car. Also the most beautiful to look at. I often leave the bonnet tilted up in my workshop, just so I can contemplate it.
5) It's English, with all that implies to a British car buff, in both sickness and in health. It has a certain sense of craft and finish in parts that most car companies blow off as too mundane to bother with, and a quietly tasteful and understated charm to many of its interior appointments. The cockpit is a wonderful place to be, and it smells like leather and old wool. It's like driving your grandfather's library.
6) There is almost nothing on the car that is not repairable by a reasonably skilled home mechanic with a good shop manual—or access to a machine shop. No black boxes. This is a car than can be kept running forever by one person. No diagnostic computers will ever have sway.
7) The Jag is perfectly poised between the modern and vintage eras. It has disc brakes, fine handling, independent rear suspension, a twincam engine and plenty of performance, yet is just old enough to evoke the Golden Age of sports cars, when Hawthorn and Collins and Moss ruled the world in cars that were British Racing Green.

So, I have to say the car agrees with me. It feels entirely compatible with how I think a GT car should operate on the road. Behind the wheel, I am somehow at home—or have come home at last.

I wasn't sure this would be the case during the three years I was redoing the engine, brakes, suspension, clutch, U-joints, cooling system and so on, with the entire front subframe disassembled and piled in parts around the garage. There was one bad night, two winters ago, when I went out to the workshop, looked around at the mountains of parts to paint, replate, rebush and reassemble, and I simply shut out the light and went back to the house.

"What's wrong?" Barb asked. "I thought you were going to work on the car all evening?"

I sat down in a living room chair and sighed deeply. "If I could arrange for God to lift the roof off my workshop and reach in with a giant shop-vac and vacuum that car and all of its parts out of my life and into the cosmos, I'd do it in a minute. I am worn out."

But then I had a big cup of coffee and went back to work.

I think it was the exquisiteness of the pieces that kept me going, and just the looks of that body shape; the will to see it in once piece again and back on the road. In many other restorations I've done, I have lost respect for the car on some level and have quietly vowed to sell it not long after finishing the project. Always on to the next big thing. But this didn't happen with the E-Type. I knew if I ever finished it, I would want to keep the car.

And the will to move on hasn't hit me yet, either.

I have discovered in late middle age there is no one thing I have to have any more to be happy. Under need or duress or changing financial conditions, my heart cannot be broken—or even badly damaged—by the loss of some material thing. There are always cheaper replacements for anything, just as enjoyable and intriguing. The world is full of fun old cars that are virtually free for the asking—or hauling away.

But I am protective of the Jaguar. If someone were to hit it in a parking lot and do serious damage, or total out the car, I would be a very sad guy, indeed. The only other thing I own that's like this is my old walnut-colored Gibson ES335 electric guitar.

They are only manufactured objects, but their like will never be seen again. They do things for us that could only come from the good intentions of the people who designed and built them. They are repositories of accumulated wisdom and human brightness, from an especially nice era of design.

If Sir William Lyons were still alive, I'd write him a letter and say, "Nice work."

April 2004

The Mysterious Case of the Huntingdon Gumption Trap

Big news: Rush Limbaugh and I are both off drugs! Yes! He's out of rehab (humble and introspective as ever), and I just finished off six months of really fun drug treatment for hepatitis C, most of which time was spent glued to the sofa—with Interferon-based super glue—watching the sun rise and set through the branches of the sycamore tree outside our window. Right before my eyes, the leaves budded out in slow-motion, turned deep green, changed to gold and then fell to the ground, and here we are in winter. Done.

During those idle months, many of my friends, knowing how much I like to read, offered me stacks of books to while away the time. Unfortunately, I had to decline, pleading low energy and a shortened attention span. I was barely able to make it through the larger newspaper headlines (WAR!) and all six pages of the used-car classifieds without giving up and taking another nap.

Yet, strangely, I managed to reread two favorite old books, a little at a time.

Why just these two books?

Because both were readable in short takes, clearly written and intriguing enough to transport one away from the less appealing present. They were *The Complete Sherlock Holmes* by Arthur Conan Doyle, and *Zen and the Art of Motorcycle Maintenance*, by Robert M. Pirsig.

Both these books have remained for decades on my small, irreducible shelf of literary "keepers" (Mayflower has moved them at least four times), but I had forgotten how good they are.

The Pirsig reread has also proved to be quite timely, because his theme of Quality is illustrated, throughout, with the theory and practice of repairing one's own machine—in his case, an old motorcycle.

But it's a great book for any mechanic to read. And, after a six-month hiatus, I am finally back in the garage, spinning wrenches and blowing the dust off the scattered pieces of my current restoration project, a 1971 Lola T-204 Formula Ford. Mechanic-ing again.

It feels good.

Of all the things I do in life, I'm probably most at home when sliding open the wide drawer of a toolbox and gazing upon a row of wrenches or screwdrivers before making a selection, while a godforsaken project (preferably British) lurks in the background. The very motion causes a pleasant electrical flow through the brain, or some release of magical endorphins. Which is probably why I spent nearly the whole decade of my 20s as a professional car mechanic. It's an addiction.

But all is not Elysian bliss. There are plenty of frustrations to mechanical work too, and Zen does a good job of delineating them for us. Now that I'm back working on the Lola, of course, I'm running into all of them.

One problem every mechanic encounters is what Pirsig calls the condition of "stuckness."

Stuckness is when you can't move forward because you can't diagnose the problem, or don't have the right tools or parts to proceed. For instance, you've broken off an engine bolt during a roadside repair, but don't have a drill or an E-Z Out to remove the broken stub from its hole, nor a bolt to replace it if you could. Meanwhile, oil is leaking out on the road. You are stuck.

Stuckness, Pirsig tells us, sometimes inspires inventive and original solutions, but may also metamorphose into something he calls a "gumption trap."

A gumption trap is any hurdle, psychological or physical, that stops you in your tracks and drains you of the energy or enthusiasm to continue. Getting the wrong part from an auto parts store and being too tired to drive back into town for the right one is a typical gumption trap. The project comes to a stop. You turn out the garage lights and go into the house to watch TV.

Discovering you've left out a critical part—say, a crankshaft thrust bearing during an engine rebuild—is another gumption trap. Your freshly rebuilt engine has to be taken apart all over again, but you are too discouraged or depressed to do it for now. Your course of progress is deflected and enthusiasm deflated like a tire with a nail in it. Lights out time again.

My good friend Bruce Livermore suffered a classic example a few years ago. He did a beautiful, first-rate rebuild on a 1275 Austin engine for his Bugeye Sprite, but left the nearly completed engine on the workbench for a couple of weeks while he was off racing his Formula Vee. He returned to bolt on the last few parts—the rear oil gallery plugs and back plate—only to notice a mud wasp crawling out of the unplugged main oil gallery hole in the engine block.

"A gumption trap is any hurdle, psychological or physical, that stops you in your tracks and drains you of the energy or enthusiasm to continue."

Were there other wasps in there? Had they deposited nesting mud? No way to know without disassembling the entire engine, hot-tanking the block again and blowing out the passages.

The engine is still sitting right where he left it, about five years ago.

His gumption came skidding to a stop, but he swears he'll redo the engine "eventually," when he's not so busy with the race car.

I understand perfectly. Bruce's Bugeye used to be mine, after all, while I was racing my first Lola 204 in the mid-1970s. Time and energy are finite.

And, now that I'm re-attacking this "new" Lola project, I'm again reminded that restoring an old basket-case race car is nothing but a long, unbroken series of gumption traps in which stuckness must be continually overcome. Here are a few of the classic traps encountered while working on the Lola recently:

1) The store is closed. The other night I planned to rivet the front aluminum bulkhead to the frame, but had only nine of the 12

identical rivets I needed to prevent an unsightly, unaesthetic mismatch of rivet-head styles. The hardware stores were closed for the night. End of panel-riveting project for the evening.

2) You have three matching bolts when you need four. I'm missing half the Girling brake caliper mounting bolts required for the Lola, and can't find them anywhere. They are specially hardened, shouldered bolts just for this purpose. So now the front calipers are sitting over in the corner while I await inspiration.
3) You forgot to send just one part to the metal platers. I tried to install the shift linkage on the Lola this week and realized I'd neglected to replate one of the two chromed shift-rod bushings. It has the patina of an iron bolt dredged from a Phoenician shipwreck, and you can't bolt a crusty old part like that on a newly restored car. Two-to-three week hold on shift-linkage installation.
4) Wrong parts. The racing fabrication shop that pressed in the new bearings on my front hubs used the wrong bearing spacers, so the brake rotors hit the calipers. They are fixing the problem, but it cost me two more full days of driving to Illinois and back, and several weeks' delay. Luckily, I don't have the caliper bolts anyway.
5) You forgot to plan. (I almost said "plan ahead," but this is redundant, as there is really no other direction in which to plan.) With the Lola, I powder-coated the rebuilt frame, but forgot to weld the radiator brackets on. Now I'll have to scorch that beautiful, pristine gray powder-coat with a stinky old welder. This sounds like a job for later. Much later, when I've calmed down.

Fortunately, our friend Mr. Pirsig has included in his book several suggestions on how to calm down and pull yourself back from the project until you can rebuild your enthusiasm. He notes, rightly, "Impatience is the first reaction against a setback and can soon turn to anger if you're not careful."

One strategy, he says, is to go take a nap, or go get a cup of coffee. Another is to sweep the floor and put your tools away. Part of the underlying frustration in mechanical work can come from not being able to find a tool. (If I'm working alone, I usually shout out loud, "It was just in my hand!") Reorganize, and start over with a clean shop. It makes problems look easier to solve.

Another invigorating trick I've used lately is to shut out the garage lights and go in the house to read another Sherlock Holmes mystery. A few hours of immersion in a gas-lit world where there are only steam trains and horse-drawn carriages is good for the soul, and helps us forget the present technical quandary.

By way of diversion, for instance, I just reread *The Hound of the Baskervilles* last week. It's the stirring tale of an English curse and a scruffy, high-performance dog whose heedless owner gets stuck in quicksand forever.

May 2004

On Being Favourably Disposed Toward the English Car

"If you aren't careful, your garage is going to turn into the British Museum," my friend Paul Roberts warned at a party the other evening, as he took yet another swig of something called "Fat Squirrel," one of our better local micro-beers.

"Look," he continued, "you've got two British motorcycles, an old Triumph 500 from the '60s and a brand-new Bonneville; a Jaguar E-Type and a Lola 204 Formula Ford. You've even got a British bicycle in your workshop—that old black Triumph 3-speed from the '60s."

"Two British bicycles," I corrected him. "I still have my 1971 Raleigh Record 10-speed as well."

"Right!" he said triumphantly, "and now you tell me you're driving out to New Jersey next week to look at an old Lotus Elan. Is there a theme here?"

I stared reflectively down into my glass of Guinness (my favorite kind of reflection) and nodded solemnly, like a person confronting some deep personal flaw. "I suppose there has always been a theme there," I admitted.

When I got home that evening, I went out to my garage, looked around and saw that Paul was right.

I had cars from Coventry and Huntingdon, motorcycles from Meriden and Hinkley and two old bicycles from Nottingham.

It gets worse.

There was a large Union Jack on the wall, a framed photo of Mike

Hawthorn standing next to a D-Type at Le Mans, an Isle of Man poster from 1982, a photo of the Vincent Black Shadow I used to own and a large poster of the Rolling Stones, circa 1964, plastered on the wall behind the drum kit in the carpeted "blues band corner" of the workshop.

I looked at our band equipment and realized that, just the day before, I'd been in a music store in Madison, looking at a new version of the legendary AC-30 guitar amplifier made by Vox. In England. That amp was the patented sound of the original British Invasion—Beatles, Stones, Kinks, etc.

So, yes, you could say there was a certain creeping Englishness going on in this workshop.

The only intrusions on this apparent thematic unity, at that moment, were an Italian Bianchi racing bicycle and a Ducati motorcycle. Also, there was an Italian flag on the wall, and a large Texas flag in the band corner, in honor of all the good music that has come our way from that state. But most of the symbolic cues in the place were British.

"Not everything from England is tasteful, of course. Look at the Wolseley Hornet. Also, England is home to more really bad wallpaper than anywhere I've ever been."

I was raised in a family that had no special regard for English ways. The Egans were all Irish, and my mother's father, Wilhelm Kroneman, emigrated here from Stuttgart in 1910 (a decision that allowed him to emerge as the sole survivor of his old high school class in Germany when the Great War was over).

Only his wife, nee Esther Bates, was of English ancestry, though she betrayed few symptoms of it, other than resolute Episcopalianism. And my father, who believed that all pretensions of privilege and rank were un-Christian, found the English class system "insufferable."

In other words, I was not exactly groomed around the family

hearth to be a future MG or Jaguar enthusiast. But the English propaganda machine went to work on me pretty early in life.

If you grew up American in the '50s, as I did, much of your view of other nations—and individuals—was colored by their behavior during World War II. And here, of course, the British shone. They were our resolute and fearless allies. To this day, I can't watch a documentary on the London Blitz without getting just a bit choked up over the sheer pluck and tenacity of the British people against terrible odds. Winston Churchill and the RAF pilots who flew in the Battle of Britain still stand as powerful symbols of the never-ending struggle to save Civilization from darkness.

And, if you were a young flying buff, it didn't hurt, of course, that the British had really great-looking airplanes. Actually, every side in WWII had rakish and handsome aircraft, but the Spitfire was one of my favorite fighter designs, right in there with the P-51 Mustang. Both used the Rolls-Royce Merlin V-12—the best-looking lump ever put in a fighter.

And, speaking of Merlin, there was the element of our shared romantic literature. I grew up on *The Wind in the Willows*, the stories of Dickens, the novels of Sir Walter Scott, tales of Robin Hood and the legend of King Arthur and the Knights of the Round Table. As a young kid, I inhabited two basic imaginary universes: The Old West and Sherwood Forest. No wonder my first bicycle, a 3-speed "English racer," was made in Nottingham.

Later, when I was about 12, my German grandfather lent me his favorite books—*The Complete Sherlock Holmes*, by A. Conan Doyle, in two volumes. I fell into these like a small prehistoric rodent visiting the tar pits at La Brea, and accidentally absorbed into my bloodstream the esthetic and cultural sensibilities of Victorian and Edwardian England, which I've not entirely lost. There's still a calabash pipe on my shelf, next to those same Holmes books.

So we've got a lot of common history, mythology and literature here as background. But it was the machines themselves that really put me over.

To put it as simply as possible, the best British cars, motorcycles and airplanes really look good to me.

I won't say they are the epitome of good taste—as that would suggest (laughably) that I possess this elusive quality myself—but they are certainly agreeable to my own peculiar biases. If I were to build Egan's Museum of Great-Looking Machines, at least half the content—and probably more—would be British. I can hardly look upon a D-Type Jaguar, an Aston Martin DBR1, an MG-TC, an AC Cobra or a Lotus Elite, Elan, 23B or Seven without getting slightly weak in the knees and losing all sense of economic discipline.

They seem to be cars that came out just right. They have what I would call flamboyance without artifice, aggressiveness without swagger, beauty without glitz. They radiate understated excellence. And to a non-flashy, reticent person such as myself (whose virtues are nothing if not understated), this means a lot.

Not everything from England is tasteful, of course. Look at the Wolseley Hornet. Also, England is home to more really bad wallpaper than anywhere I've ever been. But when they get it right, it's very nice indeed.

Another pleasant aspect to British cars is their affordability. While I have always admired Porsches and Ferraris, it's been mostly British sports cars I could afford at various times in my life. Without Sprites, MGs and Triumphs, I probably still would have been saving for my first sports car at the age of 30. Instead, I could already look back at more than a decade of sporting adventures, sudden electrical fires and road racing. Only Alfas offered comparable panache for the money, and they were less widely available.

Also, British cars have always been machines you could work on yourself. They seem to have been made for the home hobbyist and amateur mechanic (exclusively, critics would say). My E-Type is a fairly complex car, but there is almost nothing on it that I can't rebuild or repair myself. It invites—nay, demands—fiddling.

Another endearing trait of British cars is their slightly offbeat sense of humor (humour?). They come from the same country that brought you Beyond the Fringe, Monty Python and Rowan Atkinson. Some cars, such as the Bugeye Sprite and the Mini, have an endearingly cheeky and fun-loving look to them, but even the larger and more expensive cars have a dry, Graham-Hill-like drollness mixed in with all their dash and dignity.

It's hard to take yourself too seriously in a car that might stop running at any moment, or lose a wheel. As a result, British car buffs have always been a little less obsessive than those who pursue excellence and cold perfection for its own sake, and a little more likely to laugh at themselves. They have to, before someone else does.

But there's probably something deeper and less definable going on here, too.

I hate to sound like some wacky disciple of Herbert Spencer, but I have often wondered if there might also be a genetic side to the appeal that British cars hold for some of us, something innate, built into our own racial memory.

I read not long ago that modern DNA testing is revealing a much higher level of Celtic ancestry among the general British population than previously expected. In other words, not all Celts moved west or north when the Angles and Saxons moved in. Some stayed home and intermarried, or just fooled around.

This may explain why the cars themselves seem to be a romantic mixture of Saxon technology and slightly off-kilter Celtic art. I've long suspected that my Jaguar shop manual is distantly related to the *Book of Kells*. Especially the wiring diagram.

June 2004

Fetching the Lotus

"I wouldn't even sell a car to someone who had a trailer that old and dirty," Barb said on a late-night visit to my workshop.

I extracted my head from a fender well and stood up with my hands held slightly outward, like the statue of a welcoming saint—a saint with big gobs of fresh wheel-bearing grease in the palms of his hands—and appraised the trailer.

"Well," I said defensively, "you haven't seen the car we're picking up in New Jersey. It's been in a barn for the past 20 years, just like this trailer. I think the car owner will understand."

In truth, I was a little unsettled by Barb's observation, as I'd already spent three days getting the trailer ready for its upcoming trip. I'd borrowed this rig from Chris Beebe, my old friend and neighbor, who warned me it was buried under a pile of sheet metal in his old tobacco barn and frozen to the ground on two flat tires.

So I'd gone over one frigid winter morning, dressed in my best insulated Carhartt coveralls and Elmer Fudd cap, toting a compressed air tank and a scissor jack. A few hours later, with no more suffering than you'd normally experience having a molar extracted, I had the trailer wrenched from the frozen earth and sitting in the farmyard on two semi-inflated old tires.

I stood and looked at the trailer, out of breath and completely exhausted. Blood dripped into the snow, where I'd cut my hands on the rough edges of the trailer fenders.

"The things we do for cars," I said, my steaming words carried away by the subzero Wisconsin wind.

Then I deftly smashed my left hand with a breaker bar while installing a ball hitch on my blue Ford Econoline van and hopped around on one foot to facilitate cursing before hooking up the trailer. A local tire shop installed two of the best tires money can buy for $36 each, after which I towed the trailer home to install new wheel bearings. I was just finishing this job when Barb made her offhand comment about the trailer.

"I had the feeling we were driving away with a family heirloom...."

She was right. It did look a little rough.

Had to use it, though. It was a matter of history and symbolic importance.

Chris Beebe had used this trailer to haul his newly purchased Lotus Super Seven race car back from California in 1971. It was originally a small U-Haul flatbed, but Chris had spent several days with an arc welder in a friend's driveway, artfully lightening, gusseting and reengineering the trailer to fit his Lotus Seven for the trip home.

The results were structurally rather elegant. The trailer was light, low and minimalist in design, no bigger or smaller than it had to be to haul a Super Seven. It was the Lotus of trailers.

Chris used it for years to haul his very competitive Seven to SCCA races, and even took it to the Runoffs in Road Atlanta one year. The old thing had a proud history.

And now it was out of retirement, about to go all the way to the East Coast to pick up another Lotus of similar dimensions and weight.

An Elan this time.

Yes, after owning three Lotus Sevens in my life, I was finally stepping up one generation in the Lotus family and buying an Elan. (By the time I'm 90, I'll probably have a Europa.)

This whole deal came about because I shot my mouth off in print a few months ago (or shot my word processor off...whatever) in a column in which I confessed that a Lotus Elan was on my short list of favorite cars.

Lo and behold, I got an e-mail from a man named Tom Cochran in rural New Jersey. He was the original owner of a 1964 Series I Elan, bought new from Cox & Pulver on Madison Avenue, in downtown Manhattan. He'd driven it regularly for nearly 20 years, taking frequent ski trips to Vermont, and had then put it up on blocks in his barn in 1983. The fuel lines were leaking, he said. So were the rear brake lines. The replaceable subframe was cracked and rusty, and the car would need a complete restoration.

So I naturally called Tom to get more details about the Lotus and to find out how much he wanted for it. The price he named was just low enough to make the car irresistible to an excitable person such as myself, while leaving financial room for a decent restoration, at my usual free labor rate.

He sent me photos of the Lotus. It was faded red with a black top—a very early six-taillight model—sitting in all its dusty glory, covered with tiny mouse droppings, resting on blocks in the corner of an old barn made of huge timbers. Perfect, in other words.

"I'll be out next Wednesday," I told him.

Chris not only volunteered to come along, but also welded new fenders on the trailer before we left—"just so we'll be objects of admiration virtually wherever we go." Early on a Tuesday morning, we headed out for New Jersey.

"The Elan was faded red with a black top sitting in all its dusty glory, covered with tiny mouse droppings, resting on blocks in the corner of an old barn . . . perfect, in other words."

With our new wheel bearings whirling like greasy, trouble-free Dervishes through downtown Chicago, we took I-80 across Indiana and Ohio and made it into eastern Pennsylvania the first night, after a detour around a horrendous chain-reaction pileup caused by blinding snow squalls near Lock Haven. Thirty semis and 20 cars were wrecked.

We crossed the Delaware and got to Tom's place at noon the next day. He and his wife, Tabby, have a lovely old Colonial country home built in 1750, on 15 acres of hilly, wooded land. It is believed, Tom told us, that George Washington and his troops camped on this property on their way to Trenton during the Revolutionary War.

"And 200 years later, there's a British car in the barn," Chris noted.

"Redcoats' revenge," I said. "Last tag."

True to the photos, the car sat in the corner of an old green barn, with beams of light falling Caravaggio-like on its subtle and dusty curves from a nearby window. Chris (who owns two Lotus Elans himself) helped me look it over. It was as represented; rusty subframe in need of replacement, minor body damage in front of the right door, properly repaired; paint with lots of patina. An essentially unmolested car with all its original parts. We bolted on the wheels, lowered the car from its blocks and loaded the Elan on the Lotus of car trailers.

After it was tied down, I noticed that Tom had grown very quiet and still. I glanced over and saw a look of great sadness in his eyes, possibly mixed with remorse.

"Are you sure you want to do this?" I asked.

"We have a lot of history with this car," he said. "I met my wife when I first had the Lotus...we took ski trips together...I have pictures of our kids sitting in it when they were little. But it needs to be restored and I'll never get around to it. The car needs a good home."

Tabby invited us in for lunch with their daughter Oakley, and I learned that Tom had been a *Road & Track* reader since 1949. He'd owned three Morgans, including a Super Sport, and had a new Lotus Elise on order. He'd driven the Elan to Watkins Glen and to SCCA sports-car races all over the East Coast.

Tabby said, "I've got to show you a Christmas stocking I made for Tom in the mid-'60s."

She left and returned with a big red felt stocking that had a hand-stitched profile of an Elan on one side and a skier on the other. One side said "LOTUS" in large letters and the other side said "SKIING."

"His two favorite things," she told me.

Now it was my turn to grow quiet and still.

After lunch we said goodbye and headed for Wisconsin. As we pulled out of the driveway and waved, Chris said, "What a wonderful family."

I nodded. "Nice people, a beautiful setting and a handsome old car in an ancient barn. It's like something out of a dream." I had the feeling we were driving away with a family heirloom, rather than just another unwanted project car.

We made it home in a day and a half. I dropped Chris off at home, unloaded the car in the snowy darkness, rolled it into my workshop, turned on the lights and cranked up the heat.

I stayed up until 3 in the morning, puttering around with the car, cleaning, vacuuming mouse nests from under the seats and just sitting back to look at it. I finally went to bed, but couldn't sleep. Too much adrenaline. So I got up early, made some coffee and went back out to the workshop to gaze upon the Elan. A light snow was falling on the trailer outside.

I sat there for a while and tried to think if anything on earth could be more fun on a dark and wintry week than going on a road trip and bringing a nice old car home to restore.

I couldn't think of anything, even after a second cup of coffee.

Retrieving a car, of course, is always the best and easiest part. The hard work of restoration comes later, like your Visa bill after a night on the town. I believe this is called deficit living.

July 2004

Restoring Today's Cars Tomorrow

Last night I drove into the big city on wintry roads for the weekly meeting of our motorcycle club, which is called the Slimey (intentionally misspelled) Crud Motorcycle Gang.

The Cruds are, essentially, a bunch of guys who, like me, are in a stage of advanced youth—50–60 years old—and have been riding motorcycles for most of their lives. The emphasis is mostly on sport bikes, with a smattering of old British iron and the occasional Harley thrown in.

Normally, we meet at a downtown Madison bar, but last night was Fat Tuesday—what Jimmy Breslin used to call an "amateur night"—and the bar was packed. So we gathered at the nearby heated garage of our President for Life, Dr. Kenneth Clark, bringing with us an ample supply of dark beer, reasonably priced wine (too reasonably, some would say), crackers, filé gumbo, hot nacho-flavored Doritos, spicy Cajun sausage, creamed herring, peanuts and a wide selection of delightfully stinky cheeses.

You are what you eat, after all, and the Slimey Cruds are nothing if not confused.

While the Cruds are ostensibly a motorcycle club, there's also a strong streak of car enthusiasm running through the pack. Nearly every member is afflicted with that barbell-shaped gene in which the car and motorcycle infatuation are just about equally balanced, held together in the middle by a thin strand of romantic impracticality and chronic insolvency.

These guys have had some interesting cars: Jaguars, Datsun 240Zs, Corvettes, old Mercedes sedans, Mazda RX-7s, posh gas-guzzlers of the '60s and '70s, Mustangs, Jeeps, Volkswagen GTis and Beetles, Saab

turbos, swanky old Cadillacs and weird diesels of every description.

Over the years, mid-1960s Harley Sportsters have shared garages with TR-4s; Triumph Bonnevilles have cohabited with MGBs and, even now, Norton Commandos and Ducatis sit next to E-Type Jaguars. Recently, befitting our advancing age, there's been a rush on sports sedans with big engines, fat tires and handling packages, and a creeping trend toward value-laden C4 Corvettes. It's a mechanically-minded, car-loving bunch.

Anyway, there I was last night in Ken's garage, sipping on a dark local stout with the opacity of drain oil and talking about an R&T comparison test from which I'd just returned.

"How did you like the BMW 530i?" my friend, Lew Terpstra, asked.

"Nice car," I said, "if you can master the iDrive system for running the radio and heating system. It drives most people crazy."

Lew nodded. "That system seems to be a point of great controversy among BMW owners." He took a reflective sip of beer and said, "Can you imagine what it's going to be like, 30 or 40 years from now, trying to restore a car like that? Or any modern car, for that matter."

This concept brought a rare, sobering silence to the group. People stared off into space, lost in a sort of catatonic private reverie. These are guys who have restored quite a few cars, after all, and have had their lives turned upside down by devices as simple as a Lucas voltage regulator. What would a busted iDrive be like?

Lew's is not a new observation, of course. The difficulty of restoring a complex car at some future date is an old problem. My friend Wayne Chapman once bought a '50s Lincoln Continental to restore and discovered that many of the ancillary controls (window lifts, etc.) were powered by a maze of leaking hydraulic and vacuum lines, all of them in need of replacement or repair.

Naturally, he did what any skilled mechanic would do and immediately sold the car to someone else.

Along the same lines, I was once advised by a fine woodworker named Russell Beebe to steer clear of a 1930 Rolls-Royce 20/25 because the wood framing in the body was too complex for an ordinary person (i.e., me) to handle, and many of the mechanical bits were specialized and hard to find. So you don't necessarily have to get into the modern computer chip/black-box era to have trouble with the restorability of cars.

Cars that were complicated 40 or 70 years ago are still hard to work on, and those that were simple are still easy. Which is why a Model A Ford is a more inviting restoration prospect than that Rolls, and a Bugeye Sprite is more fun to restore than the E-Type I just finished. Not necessarily more rewarding, but certainly easier.

Essentially, the original buyers of those old cars paid for the added complexity with cash, years ago when the cars were new, and now you get to pay for it all over again with hours of labor and busted knuckles.

It's all tied in with the conservation of energy. The energy invested in craftsmanship leaks off into space as a car gets old and must be replaced by yet another mechanic or artisan who consumes x number of calories every day and does the job over again. Complexity is simply a larger storage battery that takes more recharging.

How's that for a new Law of the Universe? You'd never guess I got a D- in college physics and had to drop out of pre-med.

But if fancy cars have always been harder to restore, I must admit that the many new "systems" on current cars are likely to exacerbate the problem for future restorers.

I just spent the past two weeks disassembling my newly acquired 1964 Lotus Elan, and the body is now off the frame, sitting next to it in the garage. The Elan contains a great deal of very clever engineering, but it's still a surprisingly simple car to work on. There's very little of it I can't fix myself.

But what if it had traction- and yaw-control systems, with speed sensors at each wheel, a computer under the hood to analyze incoming wheel speed data and a set of pumps, working within the brake system, to apply brake pressure to the four discs individually, and another computer to monitor engine output and restrict it for best traction? Would I even think of restoring the car in my own garage?

Not bloody likely.

The Elan had a crude little added-on Motorola AM radio bolted to the underside of the dash, with one power wire, one ground wire and an antenna lead. I removed it in about two minutes. The heater controls consist of a dash-activated cable running to a lever on a water valve and a fan switch, with manual flaps to open or shut the cockpit heat. I could have easily repaired or replaced either one of these with simple hand tools by the time I was about 12 years old.

But if the Lotus had a multi-function screen with a list of menus for different radio and heat settings, or—worse yet—a GPS navigation system and CD player built in, would I even consider removing the dash and center console, as I did with the Elan? Would I want to install a new wiring harness, as I plan to do with the Lotus?

Don't think so.

Maybe somebody will be able to repair these things some day, but it won't be me.

In modern cars, of course, nobody repairs these components anyway. Mechanics simply install new parts as complete modules and throw the old ones away, or recycle them. No one pulls the back off the menu screen on a GPS system and gets in there with a soldering iron to fix the thing. It simply gets tossed.

So, for those future restorers of our present-day cars, a ready supply of superannuated parts may be the only ticket. Engines can still be rebuilt, bodies can still be sanded and painted the old way, but the electrical parts and tricky bits will just be plugged in wholesale. The vintage-parts business, as we know it now, will simply become the vintage "chunks" business.

"There's your new center control console, sir. That'll be $3700, please."

You may notice here that virtually all the parts that put the fear of God into me are electrical in nature. That's partly because I'm a confirmed electrophobe—I'd much rather remove a cylinder head or do an engine rebuild than track down the cause of a flickering taillight—but it's also because those are the parts that seem to cause the most trouble in modern cars.

Look at any of the current automotive consumer surveys and you'll find that most complaints are directed at crank sensors, engine-management computers, faulty diagnostic warning lights, failed electric window lifts, and so on. It's the electrons that let us down these days, not the rod bearings. If they're a source of trouble now, what will they be like in 40 years?

Well, they'll probably be in a lot better shape than I will, at the age of 96, but that's beside the point.

Or maybe it isn't. This is one problem I may never have to deal with.

Meanwhile, I guess I'll just pursue a best-of-both-worlds strategy and continue to enjoy the many comforts and technical advantages of newer cars (however transitory) while restricting my restoration projects—as usual—to cars made before about 1975. The rebuilding of rusty old yaw-control systems and the rewiring of iDrive units can be left to a future generation.

Truth be told, I'm not so much worried about cars of the future as I am grateful for the past. The electrical system on my first sports car, a Triumph TR-3, caught fire one evening in 1967 when I honked the horn. It was a disaster, but I was able to rewire the whole car—at curbside the next day—with a big spool of red doorbell wire bought from a nearby hardware store.

If the TR had been equipped with ABS, cruise control, heated seats, automatic door locks, a trip computer, airbags, door-ajar warning lights, seatbelt buzzers, a valet-key engine-management system, security alarms and a self-dimming rearview mirror, the wiring harness would still be on fire.

And I'd almost certainly need a second spool of bell wire.

August 2004

Mileage Highs and Lows

Well, I see fuel mileage is back in the news again. I just talked on the phone to my old buddy John Jaeger, who lives in Southern California, and he told me he'd just managed to spend $60 filling the pickup he uses for hauling around large chunks of the 1966 Mustang he's restoring. Or chunks of the Morgan Plus-4 drophead coupe, or the TVR or the Genie sports racer.

And you thought you were crazy.

I often kid John about getting a forklift to stack his cars, and from the silence on the other end of the line I can tell he's thinking about it.

Anyway, $60 for a fill-up is certainly a sum that will get your attention. John says premium unleaded in California is (at this writing) about $2.50 a gallon. Things are slightly better here in the Midwest at the moment—premium goes for a mere $1.95, and regular is about $1.75 a gallon. But the newspaper says to expect $3/gallon even here by the end of the summer. We'll see.

In any case, all this talk about fuel mileage and rising costs has me looking at my own small fleet of land sleds again, and what I find there is not exactly reassuring.

I have a 1997 Ford half-ton van with a 4.6-liter V-8—used only for towing or hauling motorcycles and car parts—that averages about 15 mpg. Interestingly, the van I had before that was a 1-ton Ford with a huge 426-cu.-in. V-8, and it got 13 or 14 mpg. Not much difference. At 65 mph, a barn door is a barn door.

Barb drives a 1995 Jeep Grand Cherokee with a 318-cu.-in. V-8 that gets almost exactly the same mileage as my van. Borderline dismal. But, like my van, it's at least used to do real work. Barb volunteers for a dog rescue organization, and spends most of her driving miles hauling around cages full of anxious canines and quantities of dog food that would make a longshoreman hang up his cargo hook.

I've pointed out to her that one of the attractive new-generation minivans would get much better mileage, have more room for cages, dog food and passengers without threatening to flip on its roof in a sudden maneuver.

And Barb has pointed out to me that she is "not a minivan person."

This merely reinforces my suspicion that SUVs are purchased about 50 percent for utility, while the other 50 percent is squandered on illusions of grandeur. But I wouldn't tell Barb that. As the owner/ restorer of an old E-Type that cost more than her Jeep, I am skating on thin self-delusional ice here.

And the mileage of said E-Type? It averages about 21 mpg. Not too terrible for a 1960s' supercar that's supposed to go 150 mph. On a good day, with factory tuning. But I don't worry too much about the Jag fuel mileage because it's a "special occasion" car that doesn't get driven enough to throw the world petroleum market out of balance. I bring it out of the garage with about the same sense of ceremony you might use in serving a really old cognac to your friends. And if gasoline gets any more expensive, I'll be burning Courvoisier in the Jaguar.

That leaves my much-ballyhooed 1988 Buick Park Avenue, which is the only fuel-efficient car we own. On a trip to visit my brother's family in Nashville two weeks ago, we averaged 31.1 mpg. It was all Interstate driving, with the cruise control set at 75 mph. That's just a tick better than the mileage I used to get with our Mazda Miata, and 3 mpg better than the 1968 VW Beetle we had in the '70s. That makes it the most economical car I've ever owned.

Other than that Buick, the real mileage king in my garage is a 2003 800-cc Triumph Bonneville, which is faster than any of my cars and gets about 48 mpg. I also have an old 1968 Triumph 500 that gets 50 mpg, but leaks so much oil its petroleum consumption index approximates that of a Hummer with underinflated tires, or maybe the *Exxon Valdez*. Oil in, oil out.

Two of my cars have only theoretical mileage ratings, as I haven't finished restoring them. The Lotus Elan 1600 is reported (in '60s' R&T road tests) to average 21 to 26 mpg in normal driving—not all that much better than the much larger E-Type. And Lord only knows what my Lola 204 vintage Formula Ford will consume if I ever get it back together. Most Formula Fords hit the Armco before you have a chance to check their fuel mileage.

"My 1962 Lincoln cost more to drive to the movies than we paid for two movie tickets and popcorn. With Junior Mints."

Getting away from the current "stable" and looking backward, my all-time personal mileage champion was a 1964 50-cc Honda C100 Super Cub motorbike—just like the one Steve Allen used to ride at the beginning of his old TV show. I bought this little beauty in 1977 for $75 at a garage sale and used it for a *Cycle World* story.

The idea behind the story was to take the little Honda on a 300-mile back-road trip in the company of a 10-speed Stella racing bicycle ridden by my friend John Oakey, to see which was more energy efficient.

The verdict? Well, during our first 80-mile day on the road, the Honda used 32 cents worth of gas, while John ate more than a dollar's worth of energy bars I didn't need, even though the Honda was carrying all our luggage. After I put that 32-cent slug of fuel in the tank, John stared at the Honda and said, "That's plain madness. You can't make a gas tank leak that slowly, much less run a vehicle...."

When the 303-mile trip was over, the Honda had used 1.8 gallons of gasoline at a cost of $1.13 and averaged 168.3 mpg. Meanwhile, John had consumed more than $4 worth of granola bars while riding, plus a peach and three apples. Also, I went out drinking at night, while John crashed, exhausted, into our pup tent, too tired to notice it was raining on his feet, or that there was an orange feral cat nesting between his knees.

Agricultural economists tell us it now takes at least one calorie of oil to produce one calorie of edible grain, and about 10 calories of fossil fuel to produce one calorie of processed food—like high-energy bars—so I guess the little Honda won. A victory offset, to some degree, by my added consumption of alcohol.

My second all-time fuel miser was also a Honda motorcycle I bought in college, a 1966 Super-90. I rode it 90 miles home from school after my last semester exam and filled it up at the end of my trip for 23 cents, which bought me 0.7 gal. of regular, for a consumption rate of about 128 mpg.

If we all lived in Hawaii, of course, these old Honda motorbikes might be the answer to our fuel problems. Unfortunately, I live in a climate where motorcycles are stored for about five months of the year while winter rages outside, so cars remain very much in the picture. Jeeps, even.

Looking backward in time for the worst mileage I've ever achieved, the award would have to go to my 1962 "suicide-door" Lincoln Continental, which was a lovely car but managed to suck down a gallon of premium about every 8 or 9 miles. It cost more to drive to the movies than we paid for two movie tickets and popcorn. With Junior Mints.

My Ford E-350 van could also be flogged downward into the 8-mpg range, but it could be forgiven because it was towing an 18-ft. race-car trailer into the wind. It was doing real work, rather than just looking swanky.

But now here we are, back in 2004, with fuel prices on the rise and a war raging again in a Middle Eastern oil country. So what is a cagey and prescient person like myself to buy, the next time another new or used car is called for?

Like a lot of car enthusiasts, I have two competing angels perched on my shoulders.

One always whispers in my ear, "Life is short. You love cars. Get something that's fast and fun to drive, and damn the torpedoes. Why worry? We will someday just be a thin, oil-stained layer of sediment halfway down the wall of the Grand Canyon, known to geologists as the Hydrocarboniferous Anomaly. Also, you aren't getting any younger, I notice...."

The other angel says, "Have you lost all ability to reason, all sense of your own importance in the world? Why are you burning four times as much fuel as Europeans, who seem to drive fun cars and live perfectly nice lives? Especially in Tuscany and certain parts of Provence, which have excellent wines and cheeses. Quit smogging up the planet. Scale down your vanity there, pardner, and get a life."

As a hopeless car nut, of course, I like listening to both these angels.

Why?

Well, because keeping them happy requires two different types of cars. One will be pleased if I find that Polo Green 1992–1996 6-speed Corvette I've been looking for (about 25 mpg on the highway), and the other will perhaps be satisfied if I offset the occasional use of that Corvette with a small, light car that gets 35 or 40 miles per gallon. Or more. I'll also get extra points for riding my motorcycle whenever the weather is good in the summer.

As luck would have it, these requirements fit in perfectly with my long-term car shopping and motorcycle riding plans.

How fortunate.

Also, I am relieved to report there's no third angel asking me to buy overweight, gas-guzzling, charmless vehicles that are no fun to drive. I'm not even sure there are such angels.

September 2004

Dread of Wiring

If you'd had absolutely nothing better to do the other night and had looked through the back window of my E-Type coupe, you might have imagined that a pair of grease-stained Minnetonka moccasins were sitting at the wheel, about to back the car out of the garage.

Alas, it was just yours truly, the home mechanic, lying upside down in the driver's seat with my feet in the air, head under the dash and a flashlight in my teeth, trying to track down a wiring short at the back of my speedometer.

While driving home earlier in the evening, I'd stuck my delicate gorilla-size hand under the dash with characteristic finesse while attempting to tighten the speedometer cable and had somehow caused all the lights in the instrument panel to go dark. So now I had an electrical mystery on my hands.

And, I must say, this is my least favorite kind of mystery. There's almost nothing on earth I hate more than lying upside down and gazing skyward into a shadowy rat's nest of wiring to solve an electrical problem.

I've always believed that every mechanic has a natural pecking order of preferred automotive jobs, based on a self-perception of skill level. Mine starts with engine rebuilds at the top, then moves downward through transmission and suspension work, from there descending through bodywork and paint and dropping sharply into the arbitrary world of reupholstery and headliner installation. Below that, at bottom dead center, is electrical repair. It's the thing I dread most. Electricity and I have never really gotten along.

Why should this be?

I hate to use an electrical analogy here, but it might just be the way we're wired.

I'm told that, in the world of professional food preparation, there are essentially two kinds of people: cooks and bakers. Cooks generally won't go near a flour scoop, and bakers would much rather roll out a big batch of croissant dough than sauté a bunch of shallots. As an amateur chef myself—and I do mean amateur—I clearly fall into the former category. On a cold winter day, I can get out the Julia Child cookbook and spend an entire Saturday making Boeuf Bourguignon, but I've never baked anything more complicated than a batch of Krispy Treats, which aren't actually baked.

Okay, I made a birthday cake for Barb once, but it turned out slightly asymmetrical, like the Guggenheim Museum or a hatbox sat upon by an elephant. The cake was either a daring artistic statement or architecturally flawed, depending on your point of view. At least the "Happy Birthday" lettering was easy to read from a sitting position at the table, like a tilted computer monitor, and all the candle wax ran to one side.

In any case, I've stayed away from recipes involving yeast or flour ever since. I prefer dishes that call for red wine and/or garlic sautéed in butter, with some wine left over for inspiration. I'm a cook and not a baker.

And if there are two kinds of chefs, I believe there are also two kinds of mechanics.

When I worked as a full-time foreign-car mechanic during the '70s, most of my fellow mechanics shied away from electrical work, as I did. We had just one guy in the shop, among 12 mechanics, who specialized in electrical problems. His name was Jack Buellesbach, and the name patch on his shirt said "Sparky."

Jack did all our generator and starter motor rebuilds and was called in as a consultant whenever the rest of us had electrical diagnostic problems with a car. It was the same sort of shamanistic, witch-doctor role that computer wizards have assumed in a modern office setting. He performed exorcisms of evil spirits that were beyond the control of ordinary mortals.

Even Jack's toolbox looked different. The rest of us had neat rows of Craftsman or Snap-on wrenches contained in drawers, but Jack's toolbox was draped in wires—test leads with alligator clips, jumper cables, volt/ohm meter wires, etc.—so that it looked like a lost Mayan temple, covered with vines. He may have had a human skull on a stick somewhere in his toolbox as well, but don't quote me.

". . . whenever the rest of us had electrical diagnostic problems with a car . . . He performed exorcisms of evil spirits that were beyond the control of ordinary mortals."

As a mechanic, I was able to do simple electrical repairs that were based on a kind of water-pipe sense of electrical flow, and I have also succeeded in doing all the wiring on a couple of my own race cars, after much slow deliberation, bovine staring and simian head scratching. I can hook up components conceived and built by someone else, but it's the theoretical stuff—the conceptual vision of force fields, the internal logic of tubes, transistors and so forth—that gives me trouble.

I can see that these things work, but I don't know why. This is a problem that goes all the way back into childhood.

As a young science nut, I imagined that I would either succeed Wernher von Braun in running the space program or become a Colin Chapman-like race-car driver and engineer, so I methodically set about educating myself in all the requisite technological fields.

Mechanical technology was easy for me. I'd go to the library and check out a book on steam engines, look at a cutaway drawing of a steam locomotive and understand almost instantly how it worked. Same with cars and gasoline engines. One reading of a chapter on 4-stroke or 2-stroke engines and I had it down.

The next frontier, of course, was electricity. I came home from the library one fine day clutching a book called *The Boy's Book of Basic Electronics*, or some such title, and sat down to read.

Everything went well for the first few paragraphs, but by the third page I was going all slack-jawed, with drool dripping on my shirt, and by the second chapter I was in so far over my head the Coast Guard couldn't have saved me with a team of frogmen.

The book was supposedly a simple child's primer on electricity, but when I flipped ahead a few chapters I found passages that read (I'm going on memory here): "...and Marconi naturally reasoned that if you pass a rapidly oscillating current through a fine mesh made of manganese and desiccated seaweed in a partial vacuum, while bombarding the anode pole with radio waves, the mesh will strip away every third electron, flattening the wave length and allowing musical notes to be rectified with some accuracy on your family's Philco."

Naturally.

I read that paragraph—or one very much like it—about three times, and then decided to go down the street and see what my girlfriend, Colleen, was doing.

It turned out she had found some stale old cigarettes, so we hid under the grapevine and smoked them. And that was the end of my electrical education.

After that, I made some feeble attempts at practical electronics, but nothing worked very well. I wound my own electric motor, using tenpenny nails and copper wire, but when I hooked this "motor" up to a battery, it just sat there and hummed faintly. No work was done, and the battery went dead.

Then I tried to build a "foxhole radio," like those supposedly used by the French Resistance in World War II, winding copper wire around a toilet paper tube and tuning it with a piece of pencil lead and a swiveling safety pin.

The big day came to try it out, but nary a sound intruded upon the utter and perfect silence inside my headphones. It was like listening to a seashell, but without all the racket. I might just as well have plugged the headphones into a block of wood. At least my motor had hummed.

At that point, I decided to give up on electricity and pursue my mechanical interests. I built a mini-bike with a bicycle frame and a 1.5-horse Briggs & Stratton engine and it actually worked, taking me all over the back roads of rural Wisconsin one summer. I also built a go kart, and that worked too. Being a mechanic was the life for me. Noise, smoke and action, instead of just smoke.

Moving back to the Jaguar and our present day, I finally found that electrical problem last night.

I'd knocked a spade connector off one of the speedometer lights (which are so dim as to beggar the term), thereby shorting out the circuit and blowing a fuse. The 35-amp tubular glass fuse looked as though a small firecracker had gone off inside it.

I hooked up the spade connector, replaced the fuse, reconnected the battery, hit the switch and all the dash lights lit up, as best they knew how. I took my coveralls off and washed my hands. Job accomplished. Done for the night.

Unfortunately, just as I was closing the driver's door, I noticed the interior courtesy lights were no longer shining. I operated the door switch with my finger and nothing happened. Had I knocked another wire off somewhere? Blown another fuse?

I hesitated by the open car door and stared at the dash. Should I put my coveralls back on and crawl under the instrument panel again?

Nope. Too tired.

I turned the garage lights out and walked toward the house, wondering where Colleen is now, and if she still has any of those old cigarettes.

October 2004

Phony Flourishes

We were bombing eastward across Nebraska on I-80, my friend Jim Wagula and I, returning to Wisconsin from a trail-riding adventure in the mountains of Colorado. Rocking gently behind us in my Ford van was a pair of dual-purpose motorcycles, flanked by the usual pile of duffel bags, riding gear, muddy boots and helmets.

Strong winds buffeted the van, and the western sky in my rearview mirrors looked like the end of the world, with angry, ink-black clouds roiling over the Great Plains. I was a little concerned, as 40 tornadoes had touched down on the plains states just the day before. "That sky looks like something from *Ghostbusters* or *Raiders of the Lost Ark,*" I said to Jim, falling back on scientific terminology learned in college meteorology.

Jim shrugged. He was preoccupied, frowning intently at an older, gold-colored Mercury Grand Marquis we were passing at that moment.

"Why," he suddenly asked, "do people glue vinyl tops on cars?"

I glanced over at the Grand Marquis, which indeed had a vinyl-covered roof, and then stared straight ahead at the highway for a few moments. It was one of those automotive mysteries I'd never really confronted before that moment.

"I think it's supposed to make a car look like a convertible," I replied.

"Well, it doesn't work," Jim said. "It just looks bad."

I nodded. I didn't care much for vinyl tops myself, though my objections were more structural than aesthetic. These tops, and the moldings around them, tended to trap moisture and eat rust holes in the roofs of the very kinds of hopelessly shot old sleds that so often sparked my attention in the want ads.

Back when I was looking for a 1963 Cadillac about 10 years ago, I'd rejected several prospective cars simply because their vinyl tops were bubbling and rusting around the edges. You could always remove the vinyl and repaint the car, but you didn't know what was under that stuff. It was like buying a house with tarpaper patches on the roof and water stains on the walls. Made you nervous.

In any case, I wasn't really sure if vinyl tops were supposed to make cars look like convertibles, or if they were intended to hark back to the horse-and-buggy era and imply a sort of coach-built, custom opulence. Either way, the effect fell short of its target.

"Is there anything sorrier than a big hood scoop that feeds air to nothing?"

"Funny thing," I said to Jim, "how often fake styling cues that are supposed to make a car look expensive or fast have just the opposite effect—they make it look cheap and slow. I suppose it helps sell a few cars to the unwary, but serious car buffs usually don't care for things that pretend to be something they aren't."

"For instance?" Jim inquired.

"Fake wire wheel hubcaps. They always look terrible, no matter how well-finished they are. Same with fake knock-offs. They never look as good as the real thing, and you can spot them from a mile away. No one is fooled."

This observation naturally led to an endless, time-killing discussion of other phony automotive stuff that always comes up short in the aesthetic and credibility departments. Nebraska is a big state, so we had plenty of time to think of many more examples. Here are just a few:

1) Hood scoops that don't go anywhere. Is there anything sorrier than a big hood scoop that feeds air to nothing? You peer in there and see a blank wall of metal or plastic. Dead-end ducts, whether for intake or exhaust, always leave a poor impression.
2) Racing stripes on underpowered cars. This is delicate ground,

because I personally like racing stripes on any number of cars, but I've always believed they belong on cars that deserve them. A 1966 GT350 Mustang looks great with racing stripes, but a 4-cylinder Mustang II with the same paint scheme makes you wince, sort of like a bad toupee.

3) Fake continental kits. If there isn't a real spare tire hidden under that unsightly bulge at the back of your trunklid, you may be subject to arrest by a roving squad of discerning automotive fashion police.
4) Wings and spoilers. Always iffy, though some cars actually benefit from rear wings. My old Army buddy, Joe Spata, bought a genuine 1970 Plymouth Road Runner SuperBird—the NASCAR homologation special—when we got home from Vietnam and it had a huge rear wing. This was probably a good thing. It seemed to help keep the car attached to planet Earth on those occasions when Joe would briefly accelerate to 130 mph on the Capital Drive freeway entrance ramp in Milwaukee late at night in a light rain. Other cars don't seem to need a rear wing as badly as that SuperBird did. Consult your dyno sheets or window sticker before deciding.

   Incidentally, Joe's SuperBird had a black vinyl top and a few superfluous scoops, but I didn't care. It was like the 500-pound gorilla that could have whatever it wanted.

"If your dash melts or makes sparks when you run it through the table saw, it probably isn't real wood."

5) Phony landau tops. If the front and rear halves of a car roof can't be folded down independently, it's probably better to skip the whole landau charade (named for a style of carriage originally built in Landau, Germany, according to my dictionary). A non-functional gold "hinge" glued to the side of a vinyl top seldom inspires as much admiration as the real thing. Plus, me and Jim think they look dumb, and we oughtta know.
6) Fake portholes in Buicks. Jim wanted me to put this in because he doesn't like them, but I do, even if they are just a useless decora-

tion. Shows you what he knows.

7) Attaching names of famous European races and racetracks to big wallowing American cars that don't corner very well and never won anything. Before Detroit noticed the popularity of European sports cars, names such as "LeMans," "Monaco," "Monte Carlo" or "Grand Prix" usually reflected victories won in rallies or races, or at least a suitability for racing, but that all changed in the 1960s. I remember being annoyed that Pontiac would borrow the name "GTO" for a car that hadn't been homologated for GT racing. No one cares any more, and this little controversy is lost in the mists of time, especially now that most of these Detroit cars are fondly regarded and remembered for their own peculiar virtues. But when I was in high school, these names were one step away from tires with raised white lettering that said "Indy Winner!"
8) "Personal luxury cars" with big long hoods and no interior space. This style, very big in the '70s, would have been more acceptable if the hood space had been needed to accommodate long and interesting engines, such as straight-8s or even glorious twincam sixes (like my E-Type's, don't you know), but it wasn't. I once rode from Wisconsin to Florida jammed into the cramped, narrow shelf that passed for a rear seat in a mid-'70s' Buick Regal, my knees all but folded up against my shoulders. Imagine my surprise when we checked the oil at a gas station in Georgia and I noticed about three feet of empty space between the grille and the engine. This car had room for a Merlin V-12 under the hood, while the rear seat cried out for just two more inches of knee room. By the time we hit Miami, I had a very jaundiced view of this hollow styling exercise. I mention this only by way of revenge and long-simmering resentment.
9) Tailfins. An obvious target, so I suppose I have to mention these, as they are useless on just about every car except the D-Type

Jaguar, with its headrest fin, and a couple of land speed record cars. But as a guy who happens to like old Cadillacs, I should probably just back away from the whole subject. I ridiculed them when I was 13, but now they look to me like a lively expression of good fun. Maybe the 13-year-old was right, but he's not around to argue with me any more.

10) A bud vase on the dash. Jim asked me to add this one. "A small styling detail," he noted, "but one that absolutely prevents any man from buying the car." The bud vase can be removed, of course, but your friends will always wonder what you've done with it.

11) Wood-grained plastic or metal. If your dash melts or makes sparks when you run it through the table saw, it probably isn't real wood. Also, a magnet will almost never stick to a real burled walnut dash. In defense of fake wood grain, however, most cars that use it shouldn't have perfectly good trees wasted on them anyway.

12) Fake competition-look accessories. I've spent many evenings poring over a new Mini Cooper S brochure and have come perilously close to buying one on at least three occasions. I like the exterior styling a lot, but find the interior much less pleasing and honest. My brochure says the tubular-shaped interior door handles are intended to mimic the steel tubing in a competition rollcage. But I happen to know the car has no rollcage, so I would rather have a door handle that looks like a nicely designed door handle. But that's just me.

With the passage of time, I may learn to overlook this slightly cloying detail and buy a Mini anyway. When I do, it will almost certainly be British racing green, with two white Cooper racing stripes, even though I don't intend to race the car.

Nobody's perfect.

November 2004

Gullwings and the Ghosts of Sindelfingen

"Have you ever driven a Mercedes 300SL Gullwing?" Executive Editor Doug Kott asked me over the phone a few weeks ago.

"Not really," I admitted. "I drove one around a parking lot once. Not far enough to form much of an impression—but it seemed pretty nice in first gear," I added pathetically.

For sheer excitement, it hadn't been exactly like Moss and Jenks in the Mille Miglia.

"Well," Doug continued, "Mercedes is throwing a 50th anniversary celebration in Germany for the SL-series cars. They're flying the press into Stuttgart to drive the current-generation SLs, followed by a day of touring the Black Forest with fully restored Gullwings from the factory museum."

"Sounds like fun," I said, falling back on my favorite expression of helpless fatalism.

"They're also hosting a dinner," Doug added, "with the famous *Life* magazine photographer, David Douglas Duncan, and Pablo Picasso's son, Claude, as guests of honor."

I held the phone to my ear and stared out our living room window for a moment, mulling over this apparent non sequitur.

"Interesting people," I said, "but what have they got to do with Mercedes?"

"I don't know," Doug admitted.

"Maybe Claude Picasso is a car designer," I offered.

Doug accepted this possibility silently. I think we were both trying to picture a car designed by a member of the Picasso family. The possibilities smacked of capricious good fun, but nothing quite as symmetrical as a Mercedes came to mind.

If I didn't yet know much about Claude Picasso, I was a longtime admirer of David Douglas Duncan's work. As a Marine in World War II, he'd shot combat photos in the Pacific, and his photos of the Korean War are still the shorthand by which I see that conflict in my mind. Duncan was also in Vietnam the year before I got there in 1969, and his dark images of the siege of Khe Sanh are still burned into the memories of most Americans old enough to remember.

On the lighter side, he'd also done many famous photos of Pablo Picasso at work and relaxing with his family at their home in the south of France. Maybe that was his connection with Claude. We'd see.

I flew into Stuttgart—ancestral home of my immigrant German grandfather—checked into our hotel, and the next day the Daimler-Benz folks provided our small band of American journalists with maps, route books and a gleaming row of current-generation SLs to drive. I teamed up with my old friend and colleague Robert Cumberford and we blasted off toward the famous *Schwarzwald.*

Driving through dark forests, green hills and one charming village after another, we got lost only briefly when I took a wrong Y-turn between Kasekellebankmitschmaltzonit and Rottweilermitrabiesin-schwimmenbaden because I couldn't read that many syllables on a small road sign at speed. Don't ask me how the Mercedes racing team ever found its way out of Germany and made it to Le Mans. Sextants, I suppose, or the smell of *escargot* sizzling in garlic butter.

Nevertheless, we rolled into the courtyard of our hotel at a decent hour, and there was a sharp and healthy-looking 88-year-old David Douglas Duncan, talking with his elegant wife Sheila and leaning on the fender of a black 300SL Gullwing. We all introduced ourselves, chatted for a while and finally got the whole story.

Duncan was fascinated by the Gullwing and came to Germany in 1955 to shoot a photo essay about the cars for *Sports Illustrated.* He wanted pictures that showed not just the car, but its *spirit*, so he shot some blurred-motion photos of a red 300SL snaking through the ancient streets of Sindelfingen, where the cars were built. When it came out in 1956, the story was called "Ghosts of Sindelfingen."

The folks at Daimler-Benz were so thrilled with this publicity that they gave Duncan a new 1956 Gullwing, black with red leather. He then proceeded to use it as his "Jeep," racking up some 450,000 kilometers in 40 years of photo assignments.

One of those assignments was to shoot pictures of Pablo Picasso and family at their home near Cannes. When Duncan showed up in his "Black Torpedo," as he called the Gullwing, Picasso fell in love with the car and asked to be photographed sitting in it.

Duncan offered Picasso a chance to drive the SL, but his wife, Jacqueline, quickly nipped that idea in the bud. Seems Picasso couldn't drive a car. At all. He'd tried once and had hit some trees in the front yard. But he took a ride with Duncan, and the two soon became enduring friends.

On another assignment, Duncan drove the Black Torpedo all the way to Moscow during the height of the Cold War and said to Nikita Khrushchev (through his "fearless interpreter"), "Since you run this place, how about letting me photograph the Kremlin treasures?—in color!"

"Start tomorrow," Khrushchev said.

So Duncan got free rein to shoot all the Czarist jewels, thrones, Fabergé eggs and icons in the Kremlin, the first person ever allowed to do so.

Driving home, he found there was a two-week wait for the ferry from Leningrad to Helsinki. Instead he chose to drive all the way to the Arctic Circle around the Gulf of Bothnia, down through Sweden and back to his own home in the south of France—in just three days. He stopped for some hot croissants early in the morning and delivered them to Picasso's house for breakfast, along with a gift of Russian caviar—delivered from Moscow by Gullwing.

"Did you ever have any trouble with the car?" I asked.

"Not one thing. Ever," Duncan said. "And the gas tank was so big I was able to drive all the way across Czechoslovakia—where I was not allowed to stop—without a fill-up."

Daimler-Benz rebuilt the car for him when it hit 300,000 kilometers, and when Duncan decided it was time to get a newer car in 1996, he gave the Gullwing to Picasso's son, Claude, who was a genuine, grade-A car enthusiast. And who, unlike his father, knew how to drive.

Claude is now in his late 50s and an avid vintage rallyist, and has driven the Black Torpedo in the modern Mille Miglia, the Tour de France, and many other rallies. He races a vintage Formula 2 car these days, as well.

I got to sit with Claude Picasso at dinner and found him to be, like Duncan, a great raconteur and a genuinely nice man, with a self-effacing sense of humor. He told us a hilarious story about rolling over a Mercedes 230SL on a remote road in Montana during the Around the World in 80 Days rally, getting the badly smashed car back on the road and then being ticketed minutes later by a Montana State Trooper for driving a car needing more than $250 worth of repairs.

"If you can show me where to fix this car for $250," Picasso told the cop in his slight French accent, "I will go there."

The officer was not amused. "You should have reported this accident," he said.

"It just happened, right back there," Picasso replied, "and you are the first human we've seen. Where we had the accident, there is no one around but chipmunks."

The cop gave him a $280 ticket for driving a car in bad repair and left.

After breakfast the next morning, I finally got to drive an old Mercedes SL myself. Robert Cumberford and I collected our route book and maneuvered ourselves over the high sills and into the red leather seats (yes, we were spared the always iffy 1950s' Mercedes plaid cloth) of a beautiful graphite gray Gullwing. The high sills and gullwing doors were necessitated by the car's deep steel space frame, but the famously hinged steering wheel made entry easier. The trunk was filled with a spare tire, so our day luggage went in the roomy area behind the seats.

A Mercedes engineer pointed out salient features. "Turn signal on right stalk; pull out this auxiliary fuel pump switch when the car is hot to prevent vapor lock in the fuel injection; four speeds, reverse over here; don't forget the handbrake lever. Have fun."

The 3-liter slant-6 started immediately and settled into a nice raspy mechanical idle. Normal clutch, slightly stiff but pleasingly succinct gearbox with synchro in 1st gear. On our way.

What strikes you about the 300SL after half an hour on the road is how "normal" it is for an exotic, high-performance sports car introduced (on my 6th birthday) in 1954. The 215-bhp engine has good torque and pulls smoothly from about 2000 rpm to its 6000-rpm redline; steering is light and accurate; ventilation good, despite the lack of roll-down windows.

The brake pedal is nice and firm, with good stopping power from those huge vented drums; suspension is quite civilized and fluid over bumps, yet very flat in corners and composed in big road undulations. And when you put your foot in it, the car just keeps going faster. (Old road tests show top speeds in the 136–148-mph range, depending on the rear axle ratio.) The only antique fly in the ointment is the swing-axle rear end, notoriously dicey in the rain or at the limits of traction. We didn't explore those limits, out of respect and self-preservation.

Overall, it's still a real car: daring, fast and exotic, yet useful, refined and eminently driveable on modern roads. Quite a piece of work for 1954.

And stunningly beautiful.

It's one of a small handful of truly great cars that defined glamour, speed and a return to better times in the sometimes-dreary atmosphere of postwar Europe. It helped transport us from grainy black & white to a new age of full color, which may be why the Duncans and Picassos of this world liked it so well.

December 2004

Total Jaguar Immersion

Well, this has been quite a month for the big cats from Coventry. I feel as if I've been living in E-Type Jaguars and just getting out for meals and coffee. Not a bad thing, mind you, just a little unusual.

It all started about three weeks ago when Jaguar press officer Anne Clinard called me and said, "Donovan Motorcar Service is going to have some E-Type racing cars we sponsor at Road America for the Brian Redman International Challenge vintage races. How would you like to drive one?"

I tried not to say the word "gosh," but I did anyway.

"Gosh," I said, "I'd love to, but my wife Barbara and I are just leaving for Canada with our own 1967 E-Type coupe. We won't be back until next Thursday."

"Perfect," she said. "You could drive over to Elkhart Lake the day you get home. It might be fun to step right out of your own street Jaguar and into our 1963 E-Type racing car."

My mind raced. Something it doesn't do easily, what with my electronic IQ limiter set at an artificially low rpm.

"Sounds like fun," I said after a moment of wondrous thought, "but I really don't think you'd want me to be driving one of your beautiful team cars. I haven't raced anything for five years, and I don't even have a current physical. Also, a race-prepped E-Type is kind of the deep end of the pool...."

"Oh, don't worry about that," Anne said. "You'd be driving our guest car, which is less powerful than our two regular team cars. Maybe you could get a physical before the event."

Dr. Richard Hill to the rescue. He kindly got me in the next morning and determined, conclusively, that not only was I still breathing, but surprisingly fit—"for a 56-year-old."

Dr. Hill, incidentally, is 57. I happen to know this because he grew up in my hometown of Elroy, Wisconsin, about three blocks from our house. My lifelong friend and lawyer, Pat Donnelly, grew up next door to the Hills. If this little neighborhood had produced a decent stockbroker, I probably could have retired by now.

But enough bitter regret; back to Jaguars.

I mailed my physical to Road America on our way out of town, and Barb and I then proceeded to drive our Jaguar 2000 miles into Canada and back. We skidded into our driveway, washed the Jag, reloaded it with racing gear and headed for Road America.

Arriving at the track, we pulled up next to three lovely silver and green E-Type roadsters under the awning of Donovan Motorcar Service, from Lenox, Massachusetts. There I met team owner Brian Donovan and two mechanics who help out with the cars, Dave Hathaway and Dane Holland, as well as the team's two "real" drivers, Bob Hebert and his son Art.

Bob was a Lotus U.S. team driver in the 1960s who has since driven in the Daytona 24 Hours no fewer than six times. His son Art started racing only in 1996, driving vintage production and formula cars, but he's won more than 60 percent of the races he's entered. A pair of naturals, in other words, and a good study in genetics.

Brian Donovan showed me the cars.

The fastest and lightest of the group was No. 61, a 1967 roadster originally built for SCCA racing by Joe Huffaker. It has a subtly downturned bonnet for reduced lift at speed, a racing T10 4-speed gearbox, 4.2-liter dry-sump engine with an output of 415 bhp. This car would be running in the very fast Group 6 class, up against big-block Corvettes and other earth-shakers.

Second down on the performance pecking order was No. 62, a 1962 roadster in a similar state of engine tune, but with fewer trick chassis parts and a more stock appearance. This car would be running in Group 2 along with my car, which was:

The No. 63 "show car." This 1963 roadster, Brian informed me, was the least modified of the group, but it still had high-compression pistons (12:1), three Weber 45 DCOE carbs, Crower steel rods, lumpy cams, racing exhaust headers and an aluminum flywheel, for an output of 350 bhp from its 4.2-liter wet-sump engine.

The chassis used heavier torsion bars and rear springs, Carrera shock absorbers and reinforced rear suspension arms. The brakes were from a V-12 E-Type, though unboosted, and a cooling fan on the drive shaft kept the rear end cool. Wheels were 15 x 7-in. Minilites, with Hoosier racing tires. The gearbox was a stock Jaguar unit.

Okay, then. Not exactly like my street car in every detail.

I donned my driving gear, and Art Hebert helped belt me into the very comfortable racing seat. The car fired immediately with a loud roar from the side-pipes, and I accelerated out pit row in a surge of big, friendly horsepower. Nice engine. I tested the brakes going into Turns 1 and 3—high effort but powerful—and headed down the long straightaway toward Turn 5, the engine howling turbinelike up through the gears.

"Never in the course of human events have I had so little desire to return to the pits."

There's probably something better than the cinematic flutter of light and shadow on the hood of a race car as you speed through the wooded tunnel of trees at Road America, but at that moment I couldn't think what it might be.

My sense of bliss was short-lived, however, as better drivers in slower cars began passing me with the dialed-in confidence that comes from racing your own car all season. Art streaked by in the No. 62 car as if he'd been fired from a howitzer. Sideways. I pointed people by and tried to stay out of the way, feeling clumsy, useless and bog slow.

"I should just be taken out and shot," I said aloud inside my helmet as I missed the Turn 7 apex by about three feet.

At least the car was good. In the endless skidpad that is the Carousel, the big Jag was easy to drive and balanced. You could hang the tail out lazily and bring it back in with the steering wheel. It had a wonderful chassis, supple yet taut, like a stock E-Type with everything snubbed down a notch.

Just as I was starting to get a little more relaxed, the engine went off-song and started running on five cylinders. We changed the sparkplugs and I went out for the next session, but the engine was still rough and down on power. TR-4s were passing me on the straights, which is not the natural order of things.

Back in the paddock, we found a loose choke in one of the Webers. I breathed a sigh of relief that I hadn't over-revved and damaged the engine.

The next morning I went out with the car running great and started to feel more at home. For about six laps. Then I heard a metallic, hammering noise, which turned out later to be a missing tooth in 3rd gear. I bypassed 3rd, but by the time I got back to the pits the whole gearbox sounded like a Maytag full of Yugo parts.

The team had no spare box for this car, so my ride was over. I berated myself for clumsy downshifting, but the guys on the team (who are, hands down, the nicest bunch of people in the civilized world) said it was probably just a defect in the well-used gearbox. The truth is probably somewhere in the middle, with the magnetic needle of guilt wavering in my direction.

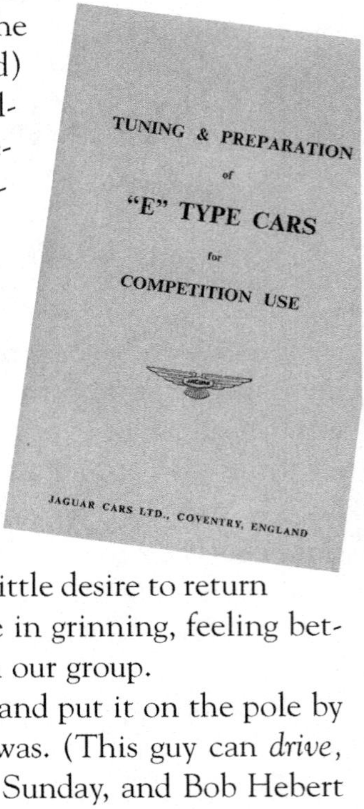
TUNING & PREPARATION
of
"E" TYPE CARS
for
COMPETITION USE

JAGUAR CARS LTD., COVENTRY, ENGLAND

Despite this, Brian offered me a practice/qualifying session in Art's No. 62 team car the next morning. I said, "Absolutely not," but he twisted my arm, so I suited up and went out "just for a lap or two."

My car was nice, but this one was spectacular—fast, smooth, balanced and thrilling to drive, with its 415 bhp. Never in the course of human events have I had so little desire to return to the pits. I did the whole session and came in grinning, feeling better because I'd qualified 5th out of 47 cars in our group.

But then Art went out in the afternoon and put it on the pole by about 5 seconds—15 seconds faster than I was. (This guy can *drive*, and so can his dad.) Art won "our" race on Sunday, and Bob Hebert was leading his displacement group with No. 61, just behind the big-block Chevy iron, when his throttle cable came loose on the last lap.

On Sunday evening, I said goodbye to the Donovan team and thanked them for putting up with me, and for the opportunity to drive again. I slid my driver's suit bag into the back of our E-Type and Barb and I headed for home, 200 miles away.

A few days later we had to leave for Barb's high school class reunion in Port Edwards, Wisconsin. Barb said, "Why don't we take the Jaguar?"

"That poor thing has had no rest," I protested. "It's overdue for an oil change and a tuneup. Still, it's running fine.... No reason not to take it, I guess."

So we drove north to the class reunion, then cruised over to Road America again on Sunday to watch our friend Bruce Livermore finish 3rd with his Formula Vee in an SCCA National, putting another 600 miles on the Jag.

We got home last night, with the car still running like a clock.

"Home," of course, is the place where you live, and I'm beginning to feel as if I've been living in E-Type Jaguars and just getting out for meals and coffee. Not a bad thing, mind you, just a little unusual.

January 2005

The New Uncivility

Last weekend, Barb and I somehow survived our first long overnight sailing trip across northern Green Bay and upper Lake Michigan. In brisk winds and 90-ft. waves (I exaggerate), we managed to navigate from our home harbor of Ellison Bay north to Escanaba, Michigan, and back. Nine hours up, and six hours back a day later.

Being relatively inexperienced sailors—this is only our fourth season—we had a few of the usual "Mr. Bean Goes Sailing" episodes involving big waves and flapping sails, but returned intact nevertheless, sunburned, happy and glad to be ashore in our snug little harbor. We fired up the communal Weber, grilled some bratwurst and had ourselves a picnic. Just before dusk, we decided to take a walk.

As we hiked along the narrow shoulder of the highway, up toward the bluffs overlooking the harbor, I noticed a disturbing trend.

More than half of the cars coming toward us moved over not at all to give us a little room, even though they had lots of empty road and no traffic coming the other way. Mirror after mirror whistled past our heads, and a couple of times we had to sidestep quickly (not to say leap theatrically, like a couple of fauns playing panpipes) to avoid skin burn from the flanks of oncoming cars and SUVs. One car came right down the shoulder at us, and we stepped back into the trees.

"What's wrong with these people?" Barb asked rhetorically. "Can't they see us walking?"

I had no answer.

Some of these drivers, of course, could be forgiven because they were on cellphones. Most people, despite what they believe, simply cannot drive and talk on a cellphone at the same time. You might as well ask them to juggle three idling chain saws and recite the complete lyrics of a 12-minute Bob Dylan song. They can't do it.

I ride motorcycles all the time, and have witnessed a recent tendency of cars to cross the centerline with depressing regularity and come straight at me, like some kind of heat-seeking missile. "Has this driver suddenly died of a heart attack?" you ask yourself while swerving toward the shoulder. Alas, no. The driver is almost always on the phone, lost in a private world of gossip, weekend plans or complex business dealings, no more in control than an ape at the wheel of an ocean liner (to steal a famous literary image from *The Magic Christian*).

So you have the cellphone factor. But others, I think, just don't care. Or, perhaps, are simply expressing the aggressiveness that defines their personalities and world view.

"We were left standing by the roadside in a small shock wave of dust and loathing."

We see it all the time here in our little rural corner of Wisconsin. Barb and I often walk our three dogs of radically different sizes after dinner on the quiet country road near our home. Even though our road is off the beaten path and is a shortcut to nowhere, I'm astounded by the number of drivers who will see two humans and three dogs dithering happily along the road and speed past without so much as a slight lift of the throttle. It happens over and over again.

About a month ago, we were walking the dogs and came across our neighbor, Emily Beebe, waiting with her small son, Graham, for the school bus to drop off his big sister, Allison, who is now in grade school. The bus dropped her off and we all stood by the road, chatting for a while. A pickup truck came around the corner and accelerated hard, straight through our midst. Kids, adults and dogs everywhere, and not a sign of slowing. Emily had to yank her two kids off the road by their arms while Barb and I waved, shouted and tried to hold back the dogs. No effect on the driver. He had places to go, things to do. We were left standing by the roadside in a small shock wave of dust and loathing.

If dark looks were laser beams, there would have been nothing left

of this truck but a smoldering shift knob.

But we, at least, were unscathed.

In the past year, we've had two joggers, one walker and a bicyclist killed by cars in our locale. And this weekend a 31-year-old mother of two small children was killed by a pickup truck while jogging on a country road three miles from our house, just around the corner from where we walk the dogs. Daylight, no traffic.

At least the driver was decent enough to call for help and wait for the ambulance. Nearly all the others have been hit-and-runs.

In any case, it's not my place to judge the man who hit the jogger; I wasn't there and terrible accidents do happen, despite our best efforts to avoid them. The terrain is hilly; the sun might have been in his eyes. Who knows? All of us can only count ourselves lucky if we get through this life without hurting someone. But, in the papers at least, the accident sounded avoidable. A little reduction in speed and a little more effort to identify with the slow-motion world beyond one's own windshield can often work wonders. As anyone who does road construction will be happy to tell you.

I should interject here, by way of confession, that I have never been an especially slow driver on the road, and when I was young, I probably drove too fast nearly all of the time. (I believe the word "maniac" was bandied about.) But, even then, it never occurred to me to speed near pedestrians, bicycles, animals or school zones—or on dates with someone's beloved daughter.

I always believed that joggers, walkers, people checking their mail, cats, dogs and family picnics in farmyards were off limits to my personal fun zone. Even my pathetic adolescent brain (as distinct from my pathetic adult one) recognized that risk is something we take upon ourselves, and no one else should be conscripted into the game. Self-destructive behavior should be just that, and nothing more. Y'all-destructive behavior is a bad thing.

Okay, enough of that. Any other gripes about my fellow motorists?

Sure. Tailgating.

Is there more of this going around, or is it just my imagination? I only ask because I am much bugged by it when returning from the city at night on my motorcycle on two-lane roads. I usually ride a little over the speed limit, even though we live in a virtual deer park, just to reduce the odds that someone will catch up and tailgate. But inevitably a car will come up behind me and drive about 20 ft. behind my motorcycle at 60–65 mph, without making any effort to pass.

Now I've got two blinding headlights in my mirrors and if I have to touch the brake lever for one of the many herds of deer that cross our roads, I'll be instantly hit from behind. And if I hit a deer, I'll enjoy the double-whammy of going down and then being run over by a car. So I usually put on my turn signal and make a right turn onto a side road, hoping all the while that the tailgater won't run me down while I'm slowing to turn. Not always a safe bet.

Another little traffic epidemic we've been struggling with, at least in the Madison, Wisconsin, vicinity, is an increase in the number of drivers running red lights and rolling through stop signs. This is not just a product of my fevered imagination; several local traffic studies have charted a large increase in these episodes. And—believe me—as a motorcyclist, I notice people who run red lights and roll through stop signs. Talk about nervous as a cockroach at a flamenco contest; I'm about two lock-up slides away from developing a serious facial tic.

Blowing red lights, of course, is as lethal to other motorists as a well-aimed artillery shell. Rolling through a stop sign is more subtle. You aren't sure if the other driver has seen you and is going to follow after you pass, or is blind as a bat and is about to make the side of his car door your final vision of planet earth. I've found eye contact with these people means a lot. Having them stop at the stop sign would mean even more.

Either way, running a busy intersection seems a monumental form of selfishness. I can't think of any other reason for it.

There, I got some of that stuff out of my system, and I hope this read (if you are still with me) hasn't been too dark and dreary. Always a risk when venting the spleen.

I guess one of the hazards of growing older is to observe gradual changes in society. Some are for the better and some are for worse, but it seems to me that a lot of American driving habits have grown worse than they were when I started driving in the early 1960s.

On the bright side, there aren't nearly as many drunks out there, but I would say there's considerably more impatient, angry or just thoughtless driving. There's been an unsettling shift to a kind of cold-blooded, "me first" outlook I don't remember as being so pervasive.

But maybe I'm wrong. Maybe American drivers aren't getting more boorish and thoughtless and are driving with about the same percentage of ineptitude and belligerence they've always had. Maybe I'm just getting older and more crotchety.

That would be nice.

February 2005

Bringing Home the Mini

"Good news, Peter! Your Mini Cooper S is in, and it's ready to go," said the British-inflected voice over the telephone last Friday morning.

Nice touch, having a salesman named Anthony Goodall from Plymouth, England, call from Milwaukee to say your Mini is ready for delivery, even if the car was engineered in Germany and styled by an American, with an engine from Brazil. The international flavor of the whole transaction was pretty heady, here in the pastoral Arcadia of rural Wisconsin.

"Great!" I said. "I'll be over this afternoon to pick it up."

Barb and I had ordered the car some six weeks earlier from the only Mini dealership in our state, International Autos in Milwaukee, which is a 90-mile drive from our home, and a rather dull flog if you take the Interstate rather than the rustic back roads.

This distance between the dealership and our house was probably one of the reasons it took me so long to make up my mind about buying the car, even though I'd spent two years poring over Mini brochures. With the service department 90 miles away, you can pretty much cross a full day off your calendar if the "check engine" light comes on. Maybe two days.

Still my five or six Mini-owning friends scattered around the country had reported almost zero problems with their cars, so I decided to go out on a limb and get the car I really wanted, rather than one that was merely easy to buy and service.

And so, on a bright sunny morning last August, Barb and I drove her forest-green Jeep Grand Cherokee to Milwaukee and ordered ourselves a new green Mini Cooper S, working on the sound principle that you can never have too many green cars.

It's been a lifelong observation of mine that interesting cars generally attract interesting sales people, and the Mini dealership seemed to bear that out. Our British-born salesman, Anthony Goodall, was a lively, helpful guy with perfect product knowledge, and it took us only about 20 minutes to get all the paperwork done. We ordered a Cooper S in British Racing Green with a white top and a pair of white Cooper racing stripes on the hood. No other options, except for cruise control and heated seats.

"Do you want a special license plate?" Mr. Goodall asked.

I looked at Barb and shrugged. "We've never done that before. Anyway, it would take me all day to think of something clever, and then later it would sound stupid. Any ideas?"

"How about BRG-1?" Goodall asked without looking up, as he filled out the car color section of the order form.

Barb and I looked at each other.

"Perfect," we said in unison.

"The car should be here in eight to 12 weeks at the latest," he said, "but it could be here before the end of September."

Six weeks later the car arrived and Anthony called me on a Friday morning. I told him I'd come right over and get it that afternoon.

But then, of course, I hung up the phone and stared blankly out the window.

What was I thinking?

How was I going to pick up a car in Milwaukee? Barb was at work, and our Jeep was all packed for a weekend trip with our friends, the Wargulas. We were supposed to pick them up after work, then drive five hours north to the Door County Peninsula on Lake Michigan, where we keep our sailboat. We were going to stay overnight on the boat, take one last end-of-the-season sail, then turn it over to the marina for winter storage. After the boat was put away, we'd get a motel, clean up and go out to dinner on Saturday night to celebrate Barb's birthday. A Farewell-to-Summer and birthday extravaganza, carefully planned.

And now the Mini was here.

I called Barb at work and explained the predicament.

"Unless you can get one of your retired friends to drive you over to Milwaukee this afternoon, we'll just have to get it some evening next week," she said.

"I hate to ask anyone to drive all that way and then come back alone," I said. "I'll think of something."

"The Cooper S was so much fun to drive, I wondered why I'd waited so long to buy one."

I had no intention, of course, of leaving the Mini in Milwaukee for another four or five days. When your new car is in, you've got to get it. Leaving it at the dealership shows you were brought up wrong.

Now, I have a brain that can barely grasp the complex logistics of grocery shopping and picking up a suit at the cleaners. Sometimes I drop off a quart of milk at the cleaners and wear my suit to the grocery store by accident. Erwin Rommel and I have nothing in common. But when it comes to collecting a desirable car (however new or old) from some distant place, a huge electrical charge surges into my brain and leaves no synapse unfired.

So, after a few moments of reflective thought, I went out to the garage and threw a motorcycle loading ramp and some tie-downs into the back of my Ford van. Then I packed for the weekend and called Barb.

"I'm taking the van to Milwaukee," I said. "I'll leave it there, pick up the Mini and drive up north to the boat and meet you guys there

later tonight. Then next week I can ride my motorcycle over to Milwaukee, load it in the van and bring it home."

"Okay..." Barb said uncertainly. "See you at the boat...I guess."

So I trundled off to Milwaukee with the van, parked in the lot behind International Autos and found our new green Mini parked in front, gleaming and ready to go. Anthony gave me a tour of the car's many features, I signed a few papers, collected the keys, shook hands and drove away into the thick of Milwaukee's rush-hour traffic.

I managed to escape the big city without getting rear-ended, despite the best efforts of a guy in a red Malibu who had NASCAR stickers all over his car and a big No. 3 decal in his rear window. I watched him move through traffic as we sped north on I-43 and he was actually drafting other cars, inches off their bumpers, while his ashen-faced companion kept the back of her head glued to the head-rest and one hand on the dash.

"Go, Dale!" I shouted, wondering if I should have ordered the Chili Red No. 37 Paddy Hopkirk Monte Carlo Rally edition of the Mini. But this could only lead to trouble.

I left the Interstate as soon as possible and took the two-lane Highway 42 along the Lake Michigan coast through the old harbor towns of Kewaunee and Algoma, across the peninsula and up the west shore to our boat in Ellison Bay at sunset. The whole shoreline was bathed in that beautiful late afternoon light you get only in autumn, tinged with chill and a hint of sadness. Most of the sailboats were already gone from the harbors. I unlocked our boat, opened the hatches, turned on the cabin lights and walked across the street to the Mink River Basin Supper Club, dining alone on the Friday night walleye special, with a brandy Manhattan to celebrate the arrival of the Mini.

The Cooper S was so much fun to drive, I wondered why I'd waited so long to buy one. It was comfortable, dead flat in corners, quick as a go kart in steering and it accelerated in silken bursts of energy. The short-throw shifter clicked effortlessly from one gear to the next, and the seat, pedals and steering wheel were perfectly placed for me.

And the car had real charm and personality. I liked looking at it and, obviously, so did other people. When you parked at a gas station, other motorists getting out of their cars positively beamed at you and went out of their way to say hello, like long lost friends. Frank Stephenson had done a good job of styling the thing—he captured the cheerful friendliness of the original Mini, but also its underlying seriousness as a performance car.

As I sat at my window table, sipping the Manhattan and looking out at the car in the parking lot, it occurred to me that the Mini covered so many bases and filled so much of the performance/aesthetic territory I carry around in my own psyche, that my penchant for collecting and restoring hopelessly shot old British cars from the '50s and '60s might be endangered. What if the new Mini was the only car I really needed?

A terrible thought, yet strangely liberating. We'd see. Addictions die hard.

Barb, Jim and Patty arrived later that night and we had a great weekend of sailing and birthday celebration. On Sunday afternoon, Barb and I drove the new Mini home and Jim and Patty followed in our Jeep.

Yesterday, I rode my Suzuki DR650 dual-sport motorcycle to Milwaukee, loaded it in the van at the dealership and came home. I took rustic back roads all the way over, riding my bike through small towns and exploring the red-and-gold tinged autumn forest roads of the scenic Kettle Moraine region. It was a beautiful autumn day with an impossibly blue Battle-of-Britain sky, filled with airplane contrails and wisps of horsetail clouds. Farm kids were carrying orange pumpkins out to their front yard stands and football teams were practicing in the fields by small high schools. By the time I got home at sunset, it had been a long day, but a good day.

I don't think there will be too much hardship, after all, in driving the Mini 90 miles for service. I keep forgetting that effort and pleasure are sometimes exactly the same thing.

March 2005

Deer in the Headlights

Call it dumb luck or poltergeist intervention, though in my case dumb luck always serves as an adequate explanation.

Last night, I was driving my Ford van down a lonely wooded lane just outside of Oxford, Wisconsin, returning from the rural home/motorcycle ranch of my friend Rob Himmelmann. It was a beautiful late autumn evening with a pale moon rising through the trees.

Suddenly, an insect about the size of a Fokker Triplane splattered against my windshield, directly in my line of sight. At first I thought somebody had thrown something at my van, but no, it was just a big old classic doodle-bug of some kind.

"Oh, good," I mumbled aloud. "Now I get to peer through a huge gob of yellow goo all the way home."

I stared at the remains of the insect cross-eyed (picture that Florida guy examining the infamous "hanging chad") and wondered what on earth it was. Unlike the aforementioned Florida, Wisconsin doesn't have many flying insects that are larger than a barn cat, mosquitoes and box elder bugs being our favored pestilence. Also, it was a cool autumn evening, without many insects flying around. None, really.

"Very odd," I said in my best Boris Karloff diction. "Perhaps I'll have to stop at an old castle and borrow some Windex."

I sighed and slowed down, planning to wipe off the bug juice with a shop rag I had in my toolbox and the $1.09 bottle of Himalayan Spring water from Chicago in the cupholder.

Just as I got on the brakes to pull over, a deer leapt into the road. I stood the van on its nose and slid to a howling stop about 20 feet short of the animal.

It was a magnificent buck with a huge rack of antlers. It stared into my headlights for a moment, then ambled haughtily off the road into the woods, and was immediately followed by four other, smaller deer. Does and fawns.

They disappeared into the trees like wraiths, as quickly as they'd materialized.

"The last 10 miles of country road to our house looks just like Kruger Game Park."

I accelerated slowly up the road about another hundred feet and three more deer ran across my path. Full stop again. I then turned right onto county highway D, and yet another spectral white-tail flickered across the edges of my headlight beams, bounding off into the hills.

This was getting ridiculous. I headed cautiously down to the Interstate and turned south toward home. On the three-lane stretches of I-road, I soon found myself uncharacteristically sticking to the middle lane, nobly letting the two outer lanes run deer interference for me. I noticed that I still hadn't cleaned the windshield, but decided not to bother. That weird solo bug had prevented all kinds of carnage and insurance claims. Perhaps I'd have its remains bronzed, difficult as that process might seem.

I hadn't always been this lucky.

About four years ago, Barb and I were driving home from a movie, when a deer ran into the road in front of us. I stopped hard and missed the deer, but, just then, a second deer launched itself out of the ditch in a graceful arc and slammed straight into the side of our car.

It died on the spot, probably from a broken neck, and never moved another muscle. Car damage was considerable—dented door and front fender, right mirror knocked off and chrome strip torn from the side of the car. We lifted the dead deer off the road, drove home and called the sheriff's office to report the accident. The officer on the phone sighed.

"Do you want the meat?" he asked.

I was slightly nonplussed by this question. I hadn't gone deer hunting since I was in high school, and was not really equipped to handle the processing of a large dead animal. Also, Barb works for about three wildlife rescue groups and was brokenhearted that we'd killed this beautiful doe. A venison dinner at our house would be a dreary thing indeed.

"No," I said, "we don't want the meat."

"Okay," he said wearily. "Well, then I guess we'll have to send someone out to pick it up."

We took the car into a body shop the next day, and the damage, as I recall, amounted to about $1600.

Interestingly, we had planned to take our motorcycle to the movie that night, but had changed our minds at the last minute because I saw lightning in the west. If we'd been on a bike, the deer would have landed in our laps.

Since that night, we've hit no more deer, but have had so many close encounters I can't begin to count them all. Here in southern Wisconsin, the deer are everywhere, and deer-car collisions have become epidemic. The *Wisconsin State Journal* reports (on the very morning I'm writing this) that crashes between vehicles and deer rose 5.8 percent last year, with 21,666 reported collisions, 792 injuries and 13 fatalities.

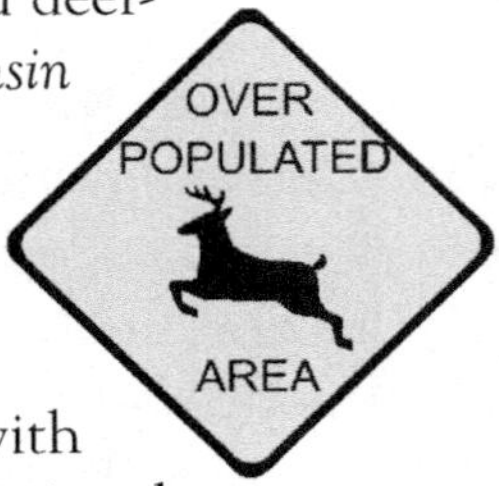

I have three good friends who have hit deer with motorcycles in just the past two years—one badly injured and the others banged up. My friend Steve Olsen destroyed his BMW R1150GS hitting a deer just two months ago, and Barb and I nearly hit one the summer before last on nearby Old Stage Road with our own motorcycle. If the deer hadn't lost its footing and stumbled while racing across the road, we would have T-boned it.

But enough figures and anecdotal evidence. We have a real deer problem. Some nights when I drive home from the city, the last 10 miles of country road to our house looks just like Kruger Game Park, with whitetail deer rather than Thompson's gazelles and wildebeest randomly crossing the road. I drive no faster than about 40 mph on these roads now, scanning the outer reaches of my headlight beams for incoming missiles. Almost no trip from town is deer-free.

It wasn't always this way.

When I was in high school during the mid-1960s, we hardly ever saw deer on the road at night, even though I lived in hilly and densely wooded farm country in central Wisconsin.

Nighttime, in fact, was when I drove fastest—and I drove too fast most of the time. There was less traffic at night on those small back roads, more warning of oncoming cars from headlights over the hill, and my teenage night vision was laser-sharp with the headlights on bright. It felt like the safest time to drive. It never occurred to me I might hit a deer. You saw one every so often, but sightings were rare and almost exotic, like spotting a great horned owl, or Bigfoot.

But now they're everywhere. So why the big increase in the deer herd?

Probably a lot of reasons. First, I don't hunt them any more—not that I ever made much of a dent in the deer herd with my meager hunting skills, anyway—but lots of other people like me no longer hunt. Or, increasingly, never did. When I was a kid, nearly everybody on farms and in the small Wisconsin towns hunted, and I think it's safe to say that *somebody* knew where virtually every legal deer in the county could be found when hunting season arrived. There wasn't much sentimentality about deer, and they represented a lot of nearly free food (and sport) at a time when most family incomes were small.

Suburban sprawl is said to be another factor. Deer thrive in suburban settings, where they're protected in small woodlots but have lots of flowers, vegetable gardens, orchards, etc., to feed on. In his great book, *Wildlife in North America*, naturalist Peter Matthiessen says there are more deer now in North America than there were at the time of Daniel Boone.

And then there are the simple economics of tourism and game management. Reduce the herd too much and there's less revenue from out-of-state hunters. Fewer licenses, less beer and pizza sold in bars, fewer motel rooms occupied. It's an industry.

But, lately, it seems like an unregulated industry. Deer, cars, motorcycles—and my van—are not a happy mix, and the numbers seem badly out of skew. Living in the country as we do, I've begun to see deer as one of the most likely threats to my own longevity, in about the same category as heart disease or cancer. They're certainly no less lethal.

I don't know what the answer is, other than trading my cars and motorcycles in on something slower, like a buckboard and a horse. I could go into town on Saturdays and buy some calico cloth, a few of them hard candies, a sack of flour and a chandelier from Paris, France.

Disparity of speed—a relatively recent phenomenon, historically speaking—is part of the problem. I'd guess the pioneers and Indians never ran into deer. Not fatally, at least. Maybe our local Amish are on to something.

Or maybe we need more predators—other than humans. Last weekend's paper said the Wisconsin wolf population has risen dramatically and is now getting slightly out of hand. Also, for the first time in my life, we can hear coyotes howling at night.

And, luckily for me, we also have an apparent increase in the number of inexplicably huge insects that come out of nowhere and smack your windshield. All a part of Nature's fine balance.

April 2005

Our Old Pal, The Prince of Darkness

My friend Jeff Zarth called me up last night and said, "What's your e-mail address? Someone sent me a whole list of Lucas jokes and I think you'd get a kick out of them."

Now, I thought I'd seen or heard just about every Lucas joke in the world, either on bumperstickers, T-shirts or by word of mouth—the warm English beer/Lucas refrigerator joke, the three-position Lucas switch with Dim, Flicker and Off, and so on—but Jeff's list contained a couple of variations I'd never heard. For instance:

"I've had a Lucas pacemaker for years and have never experienced any prob..."

And: "A guy peeked into an old Land Rover and asked the owner, 'How can you tell one switch from another at night when they're all the same?' He replied, 'It doesn't matter which one you use; nothing happens.'"

I was also pleased to see that an R&T quote made the list. It was from an old cross-country touring story I did years ago with my friend Chris Beebe, who muttered, while grinding away on a Lucas starter, "If Lucas made guns, wars would never start."

I should point out here, for younger readers who didn't live through the '50s, '60s and '70s era of British car ownership, that Lucas was the British electrical firm that manufactured most of the wiring, switches, coils, points, condensers, horns, headlights, fuses, voltage regulators, etc., that were used on nearly all British cars in those days.

And let us just say, delicately, that some of these devices did not inspire praise. Especially among those of us stranded by the roadside on a rainy night.

Or stranded in the grass by the side of the track on Turn 5 at

Road America.

Or stranded on East Washington Avenue at night in Madison, Wisconsin, while our TR-3s smoldered and filled with smoke and our dates in their prom dresses stood on the sidewalk, watching for cabs.

Or…well, you get the picture. Notice the overuse of the word, "stranded."

British cars of that era suffered from a variety of reliability problems, but any time you saw one parked by the side of the road you suspected the complicity of some Lucas product. It was like seeing a known pyromaniac standing in the crowd at a burning apartment building; he might be guilty and he might not, but his presence made you wonder.

"As our generation dies off, perhaps Lucas jokes will gradually fade away."

Anyway, this long record of trouble has spawned a lot of jokes, but not everyone has always been amused by them.

When I first came on staff at *Cycle World* magazine in the early '80s, I wrote a story that contained a typically derisive remark about the Lucas electrics on my 1967 Triumph Bonneville, and I got an indignant letter from the legal staff of Lucas in England. They explained that Lucas was now a high-tech company that manufactured the highest quality automotive and aerospace components, and that its dedicated engineers and skilled workers should not be made the brunt of these tired old jokes, which were rooted in a bygone era. The letter politely but firmly asked me to cease and desist in my derogatory comments.

I was somewhat taken aback, not to mention abashed, and realized that these folks had a point. Should young, hard-working engineers be held perpetually responsible for switches and voltage regulators that were probably designed before they were born? Maybe it was time to stop these jokes, just out of fairness. Also, the lawyers sounded kind of angry….

I showed this letter to my Managing Editor and asked him what I should do.

"Write back," he said, "and tell them that if they hadn't made junk for three decades, there wouldn't be any Lucas jokes. You can't rewrite history with a letter from your legal department."

My Managing Editor spat these words out with a certain vitriol. Perhaps because his Triumph Spitfire had just been towed that weekend. Or maybe it was because he couldn't get his AJS 500 vintage racer started because there was no spark. I can't remember which.

Anyway, he had a point. But so did Lucas. For all I know, they're making excellent stuff these days (and were then) and are no more responsible for my old Triumph Bonneville's ignition problems than I am for the failure of the Bay of Pigs invasion. Which I swear I had nothing to do with.

Also, I think Lucas was sometimes made a scapegoat for other malfunctions. I've run into quite a few former MG owners, for instance, who think the electric fuel pumps they had to whack with a knock-off hammer were made by Lucas, when in fact it was an SU pump. I wouldn't swear it didn't have Lucas parts in it—I've never taken one apart—but it was an SU product. Also, their troublesome Smiths instruments, which had the life span of a fruit fly, had nothing to do with Lucas. Unless they were dark at night.

Lucas did have some problems, though, and I've had a few. Looking back on almost four decades of British car and motorcycle ownership (27 vehicles in all), here is a sampling:

1) Lucas starters have given me a lot of trouble. In all fairness, though, it's usually been on Formula Fords, where the drive gear slides forward on the shaft and engages the flywheel ring gear under heavy braking. Still, these things wear out quickly, even in street cars. I've had two of them fail—on a TR-3 and a Sprite—and I automatically rebuild them on any "correct" British restoration project.

   But the Lucas starter really was the bane of FF racing—the weakest link in cars that were otherwise pretty bulletproof—and most people replace them now with some kind of reworked

Japanese unit. As recently as last summer, however, I helped push-start a vintage Formula B car with a bad Lucas starter at Elkhart Lake. It was like a recurring nightmare—the mechanic's version of the Missed Final Exam dream. Will we ever be free?

2) On my first H-Production Bugeye Sprite, I had an intermittent stutter-and-die running problem that cost me three DNFs. I replaced the plugs, points, coil, condenser, fuel pump, distributor, cap and plug wires before I discovered a bad connection inside my sealed Lucas battery kill switch. I was real mad at that switch, and later hit it with a big hammer in the privacy of my own garage.
3) In 1967 I honked the horn in my TR-3 and the wiring harness burned up. I've gotten a lot of mileage out of this pathetic incident on these pages, so I won't belabor it.
4) Several of the rocker switches on my two early '70s MGBs failed. Pretty junky stuff, and expensive to replace.
5) I've had one failed Lucas Sport Coil on a Lotus Seven, although these are usually pretty good, and I still use them on vintage racing cars. They put out a big jolt, and they look right.

That's about it, really. Some of the old Lucas parts work perfectly fine, if you service them once in a while, and others are actually quite beautifully crafted—the old chromed license plate and "wing" lights, for instance, add considerably to the understated elegance we like so much in old British cars. They are part of the charm. And a lot of these components are simply "of their time," and work no better or worse than their contemporary parts from other countries. Old stuff is old stuff, and you have to make allowances.

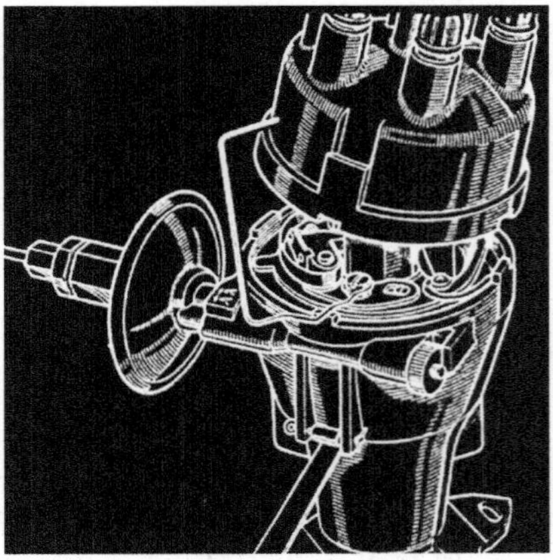

Lucas distributors, for instance, are pretty reliable if you occasionally do a proper tuneup and grease the advance weights and put some oil on the shaft. I've replaced two mysteriously malfunctioning high-tech electronic ignition systems in race cars with Lucas distributors and solved all my problems.

One of the rare benefits of these old Lucas ignition and wiring systems, in fact, is that you can usually track down problems with a simple test light or volt/ohm meter and fix them yourself. Some Lucas parts do have a short service life, but they are at least understandable, even by an electrical idiot like me.

Which is more than you can say for electrical components in modern cars.

Our local paper this week ran a story on this very subject, noting that the reliability ratings on several models of Mercedes-Benz, BMW and Volkswagen had recently dropped to a new low. The principal reason?

Electrical trouble.

Too much complexity. Servo motors, black boxes, computers, 57-way adjustable seats, information screens, nav systems, warning lights, frill units and frippery sensors. Yes, even Germans can have technical problems.

So maybe old Joseph Lucas will have the last laugh. As our generation dies off, perhaps Lucas jokes will gradually fade away, renamed for some other company that's not even British.

Or maybe PCs will take over as the new target of electrically fired derision. My home computers have caused me infinitely more frustration in recent years than the Lucas electrics in my E-Type, which have been almost trouble-free. Maybe Joe Lucas jokes will be displaced by Bill Gates jokes.

It could happen. Humor, after all, is nothing more than pain and despair, recollected in tranquility and mitigated by half-demented laughter.

May 2005

Minis, Strats and Other Instruments of Transport

As we drove north on Interstate 24, I felt elated, yet naturally paranoid at the same time.

Barb and I had just spent the Christmas holidays visiting my brother Brian and his family in Nashville, and were on our way back to Wisconsin in our new green Mini Cooper S. The roads were freshly plowed, but a recent blizzard had sent scores of cars spinning and skidding off the road and into the median. Many of them were still strewn along the shoulders, abandoned and awaiting extraction by tow trucks.

Near Fort Campbell, Kentucky—where I'd spent 10 weeks of the winter of 1969 in basic training—we passed an SUV lying on its roof down in the median, looking like an overturned toy. I looked out on the snow-covered training grounds of Fort Campbell and said to Barb, "I'm glad it wasn't snowing when I was here. We had six weeks of freezing rain, and that was bad enough."

Barb nodded. She's heard enough about Fort Campbell from me to write her own book, though I doubt she will. Basic training is an intense time in your life, and its sights and sounds get imprinted on your brain. Especially if it's about 34 degrees and raining every day, and you are preparing for a war that never ends. Dark times.

But in one of those wonderful, sudden focus shifts that life sometimes provides, I suddenly snapped out of that gloomy reverie and realized it was now Christmastime, nearly 36 years later, and I was at the wheel of the Mini, my first brand-new, really fun road car and that Barb was by my side. Hence the elation.

Further enhancing this upbeat mood was the physical presence of a guitar case nudging my elbow.

Yes, while cruising the guitar shops of Nashville, I'd accidentally bought myself another electric guitar, a 1957 reissue Fender Stratocaster in two-tone sunburst, with a maple neck. It was an instrument whose subliminal appeal could probably be traced to too many hours spent looking at the cover of a Buddy Holly album at an impressionable age. Or maybe listening to Dick Dale. Who knows? The Strat is such a deeply embedded symbol of '50s' American culture it makes the '57 Chevy, by comparison, look almost like a Communist plot.

"If they had 'guitar seats' instead of 'child seats,' that Fender would have been strapped into its own padded chair."

Anyway, I'd found this Strat, slightly used, at Gruhn's guitar shop in downtown Nashville. I bought it after a typically sleepless night of wondering if I should spend the money on a guitar I didn't really need or follow Thoreau's timeless advice to simplify your life and shun extraneous possessions.

After pondering this all night, I finally realized that they didn't even have Fender Stratocasters when Thoreau was alive. "What did that guy know about electric guitars?" I asked myself. "Nothing!"

Thoreau was lucky, really. They didn't have Jags, Lotus Elans, Ducatis, Minis, Morgans or Mustangs then, either, or he would have been a quavering mass of material lust like the rest of us. Anyone can shun 19th-century stuff that's no fun to own.

So Barb and I sat in front of Gruhn's guitar shop until it opened in the morning and I ran in and bought the Strat. It came in a big rectangular tweed case with the aura of an old steamer trunk.

Not until I walked out of the store toward the Mini, however did I pause to wonder if the guitar case would fit in our car. I opened the rear hatch and found that the guitar case would *just* fit, if I slid it part way between the front seats. So I drove away with the rear of the case touching the rear hatch glass and the front of it brushing my elbow. A vintage reissue guitar in a vintage reissue car.

That guitar was a source of quiet satisfaction during the entire trip home—as was the Mini—but it also caused the mild paranoia I mentioned earlier.

For some reason, I always think of musical instruments as being more vulnerable—and somehow more irreplaceable—than the vehicles in which they are carried. It's almost like the protective instinct of a parent for a child. If they had "guitar seats" instead of "child seats," that Fender would have been strapped into its own padded chair. I never put a guitar—or amplifier—into the trunk of any car without thinking how awful it would be if I were rear-ended in a traffic accident. It's hard to explain why the sound of splintered wood or shattering vacuum tubes should be worse than the sound of crunching sheet metal, but it is. Maybe it's the twang-and-broken-glass factor, the cartoon soundtrack for every disaster.

Or maybe it's just pure, unvarnished memory. The only real car accident I ever had, after all, occurred on my way home from picking up a Fender amplifier.

It was in the autumn of 1966, right before I went off to college. At that time, I owned a Harmony electric guitar, but no amplifier, so I convinced my parents that I should be allowed to buy one—despite my looming college expenses—using some of the money I'd made by working all summer on a railroad section crew.

They reluctantly agreed, and I borrowed my dad's 1962 Ford Fairlane 500 and drove to a famously cheap music store in West Allis, near Milwaukee, 160 miles from our home town. My buddy Pat Donnelly, who was the drummer in our garage band, went with me. There I bought a used Fender DeLuxe Reverb amplifier for $180, setting it carefully on the back seat of the Fairlane, and we headed for home.

All was well until we accelerated out of Mauston, Wisconsin, just 12 miles from home, and a farmer driving a pickup truck with five children in the cab ran a stop sign and crossed the road in front of us. He slammed on his brakes, blocking both lanes. I swerved hard right, missing the front of his truck and piling into a bunch of wood posts in the middle of a traffic island. We went from 45 mph to zero in about 10 feet.

The Fender amp in the back seat slammed against the back of my driver's seat and crashed to the footwell with a loud clanging of reverb springs.

Pat and I were both wearing seatbelts, so we didn't go through the windshield. But the steering wheel was badly bent forward by my hands, and the front of the car was smashed like a cheap accordion, hood buckled and radiator steaming. Luckily for both of us, I was an early disciple of the Giuseppe Farina school of arms-out driving and had the bench seat so far back that our heads missed the dash and steering wheel when we whipped forward.

I instantly leapt from the car, ran over to the pickup truck and shouted at the farmer, "You've wrecked my Fender amp!"

"Lord, I'm so sorry," he said, looking somewhat nonplussed by my odd response to the accident. "I had all these kids in the cab and I never saw you coming."

Then he backed up to the stop sign he'd missed and put on his right turn signal.

The deputy who investigated the accident turned out to be his brother-in-law. The farmer denied having run the stop sign—despite the big skid-and-manure marks he'd left in the road. "I was turning right, and I guess this kid panicked and thought I was crossing the road."

The deputy looked at the skid marks I'd pointed out, scratched his head and said to the farmer, "I shore hope you and Mary can come over for dinner on Sunday. Bob's gonna be there with Ellen." Then he wrote an accident report that showed me inadvertently swerving off the highway for no apparent reason. A sudden, self-destructive whim, perhaps. So much for the honesty of simple rural folk.

That tough little Fender amp was miraculously undamaged, but it was several days before I felt it politic to plug it in and play my guitar. The joy of the trip had evaporated like steam from the crumpled Ford radiator. My dad's car was ruined, and the mood was dark at home.

The only bright side to this picture was that my dad traded in his wrecked Fairlane and used the insurance money to make a down payment on a new 1966 Mustang. I actually got to drive the Mustang a few times before my insurance was cancelled because of the accident.

I sold my amp and electric guitar a year later, to help pay tuition and to buy Christmas presents.

Looking back on this era now, it sometimes seems that every triumph, however small, carried with it some corresponding disappointment or setback, some cloud of acrimony to obscure an otherwise sunny moment. I started to believe then, almost superstitiously, that the gods required some retribution for every moment of happiness.

Maybe they do; I don't know. But we made it home from this recent trip without any accidents or mishaps.

Barb and I had a wonderful time in Nashville with my brother and his wife Sarena and their four kids, including a new baby boy named Patrick. The Mini arrived in our driveway unscathed; no one has cancelled my insurance, and I love my new Strat. I played it yesterday in practice with our garage blues band, The Defenders, which still includes my old buddy Pat Donnelly.

This much condensed happiness should theoretically make a person nervous, but I'm not expecting any bad luck this time around, just so fate can balance the books. Sometimes it turns out you've already paid.

June 2005

Rebirth of the Car Worth Having

It was a warm, sunny morning in Southern California, and I was quite content to be there. I'd just flown in from wintry rural Wisconsin, where our 35-year-old Jacobsen snow-blower had been acting up, threatening to cut us off from civilization until spring. This could have been trouble, as we were almost out of sourdough and horse liniment, and the little woman was hankering for a bolt of calico cloth from the dry goods store.

Luckily, a quick sparkplug change to an ancient Champion N3, last used in a racing Bugeye Sprite 30 years ago, allowed me to get the thing started and snow-blow my way to the highway. This, in turn made it possible to drive my Mini Cooper S to the airport. Proving, I guess, that all British car culture stands on the shoulders of a previous generation. Or something.

Anyway, I made it to California, and on a warm Sunday morning drove out to visit my old sports-car racing buddy, John Jaeger.

John lives in a place called Silverado Canyon, an old silver-mining village spread out along a serpentine road that climbs into the hills. The canyon itself has a nice Early California feel. It's a mixture of old cabins and newer houses, scattered with a funky array of vintage cars—faded VW microbuses, Ford pickup trucks of the flathead era, old MGs and Triumphs, etc. John has added his own contribution to this scenic mix with a couple of Morgans, a TVR and a 1964 Buick Skylark. It's sort of a Buddy Ebsen-meets-Bob Tullius environment.

Next to John's front door is a pile of bricks. He uses these to chuck at the rattlesnakes that occasionally drop into his driveway from the rocky hillside that looms over his property.

I blow snow; he throws bricks at rattlesnakes. Every part of the country seems to have its own peculiar drawbacks.

When I got to John's house on Sunday morning, we looked at his car projects for a while, then drove down the canyon for breakfast at a little roadside café. As we dined on *huevos rancheros*, I gazed out the window and noticed a red 1964 Corvette convertible parked across the street. "Nice Corvette," I said to John.

He nodded. "It belongs to the woman who works here at the café."

"Are you still looking for a mid-'60s' Sting Ray?" I asked John. These cars had been on both our radar screens for as long as I could remember.

"We are living in an era when a whole raft of new cars are—finally—giving the old classics a run for their money."

He shook his head. "Not any more. Prices have gotten too high. Good '60s' Sting Rays are selling for $40,000 or more these days. They're nice cars, but they just aren't worth that much to me. You can buy an almost new Corvette for that—or a nice C4 convertible for about half as much," he pointed out. "I start to lose interest in old cars," he added, "when they cost more than about $20,000. Takes the fun out of it."

Strangely, John's comment distilled an idea that had lately been fermenting in my own slow-thinking brain like a rogue batch of sour mash in a nuclear reactor. Namely, we are living in an era when a whole raft of new cars are—finally—giving the old classics a run for their money.

This slightly uncomfortable realization first hit me several years ago, when I bought a really tired-out old 356B Porsche for $14,000 and then spent another $10,000 bringing it back to roadworthiness. The whole experience was a lot of fun, but, when all was said and done, I'd sunk one year of work and $24,000 into a 1963 Porsche 356B coupe. That still needed paint.

In California last week, I picked up a *Classic Car Trader* and noted with interest that Porsche Boxsters, three or four years old, can be bought for about that same $24,000. Or less. And these are *convert-*

ibles, with air conditioning, CD players, 217 bhp and no swing axles. A Boxster S (with 250 bhp) costs a bit more in the used market, but it's still cheaper than a decent 356 Cabrio or Convertible D. And a brand-new 2005 Boxster S costs slightly less than a well-restored Speedster.

So if I still had that $24,000 in the bank and wanted a Porsche, would I run out and buy another 356 Coupe, or go for a good used Boxster with low miles?

A tough decision for an old-car preservationist like myself, but I have to admit I'd be very tempted by the Boxster—which reminds me of my favorite old Porsche, the 550 Spyder.

Jaguars present a similar problem. After I'd finished restoring my 1967 E-Type coupe a couple of years ago, I couldn't help noticing that the money I'd spent would have bought a pretty nice used XK8 convertible. And among all the test cars I've driven over the years at R&T, the XK8 is one of a small handful of favorites. It's not quite as initially stunning as the E-Type, but it's a beautiful car, fast, torquey and easy to live with. It's also nearly new and the top goes down. (That convertible top in a Series 1 E-Type will set you back another $10K–$15K.)

Park the two of them side-by-side in the same used car lot, at the same price, and I'd have a real quandary on my hands. Both are magnificent, but one has air conditioning and a CD player.

Notice how I keep mentioning those two features? That's because I'm getting old.

Another push-pull combo in my own small universe of car lust is the recently arrived 2005 Mustang GT stacked up against my long-time favorite Mustang, the 1965–66 GT Fastback.

While in California last week, I got to borrow R&T's Mustang GT convertible test car for three days. I like this car a lot, and like the way it looks. It's fast and handles well, and it also makes 300 bhp. Cost? It's a little cheaper than a fully restored 1966 Mustang convertible, and quite a bit cheaper than a nice GT Fastback with the 271-bhp Hi-Po K-code engine.

Choose wisely, Weedhopper. (My Ford-buff colleague Tom Cotter says it's quite clear that any right-thinking person needs a Mustang from both eras. He may be correct.)

And then there's my own 2005 Mini Cooper S. I paid $19,900 for this car.

Just today I got my new 2005 *Sports Car Market Pocket Price Guide* in the mail from my publisher friend and fellow old-car nut, Keith Martin. Keith's booklet informs me that the current price range for a genuine original 1964–1966 Mini-Cooper S is $15,000 at the lower end and $22,500 for a concours example.

Now, I love old Minis—hardly anything on earth is more fun to drive—but you might say I've already voted with my pocketbook on this one. My brand-new "concours" Mini, for $2600 less than the original, seemed like the way to go. The extra 95 bhp, a/c and heated seats didn't hurt, either. Plus I can play my Muddy Waters CD box set.

The perceptive car buff will notice a couple of patterns, however, in all these contests between old and new.

First, the old car is always slightly lighter, "purer" and better-looking. Those mid-1960s' cars were instant classics and, for the designers who had to update them, hard acts to follow. Sort of like going on stage right after Jerry Lee Lewis or James Brown. If somebody had told me I had to "freshen up" the 1966 Mustang Fastback or redesign the 1967 Corvette Sting Ray Coupe, I would have laid my pen down on my drawing board and retired to the nearest bar. And I'd still be there.

Another notable thing about the old versions of these cars is that they are always simpler and more repairable. No black boxes, no modular anything. Mere mortals can fix whatever goes wrong. Difficult as my E-Type was to restore, I believe a worn-out XK8 will be appreciably harder to resurrect, 38 years from now.

But the new cars have a few things going for them too.

Most notably, all of them are easier to live with on the highway, with taller gearing for a more relaxed gait on the open road. I'd drive my new Mini from Wisconsin to California without a second thought, while the old Cooper S would be spinning its heart out on the Interstate—and the transmission would be worn out when you got there. Even the old Mustangs and Corvettes are pretty undergeared, unless you swap transmissions or rear ends.

The new cars also have more horsepower and get better mileage, while polluting less. (Modern Corvettes can achieve 28 mpg on the highway—about the same as the 1968 Beetle we once owned.) New cars handle better and have modern tires with improved grip. They also have real brakes and go much longer between tuneups, with engines that are almost sealed units. If modern cars could lose a few of their unwanted gizmos (like the automatic door locks on our Mini, which drive me nuts), they'd be almost perfect.

That said, I still like the spirit, simplicity and good looks of those old 1960s' classics, and I'll probably always have one under reconstruction in my workshop. It's a hobby and a way of life.

But it's nice to see genuinely appealing modern alternatives standing by in showrooms, for those occasions when the old favorites price themselves beyond our own personal limits, or, as John says, grow too expensive to be fun any more.

Back when I was a full-time car mechanic in the 1970s, watching cars get slower, uglier and duller every year, I would never have believed this could happen. We live in good times, again.

July 2005

The Cadillac from Cyberspace

In a fun-filled, often tragic, lifetime of buying, selling, restoring and racing 50-some automobiles since I bought my first car (a 1951 Buick Straight Eight) in 1964, you might say I've learned a few things.

Or you might not. The smart money says I haven't, and so does the guy who checks the gas meter in my workshop.

Nevertheless, I have come up with some hard and fast Garage Rules over the years, and any day now I'm going to have them chiseled on stone tablets in some classic script—Roman or Hebrew would look good—and mounted above my workshop doors. They are as follows:

1) Never buy any car that was disassembled by someone else.
2) Never buy a used race car that wasn't driven off the track and straight onto a trailer after doing fairly well in its last race.
3) Avoid any old road car that doesn't start and run, but "should" or "used to."
4) Don't, under any circumstances, buy a car you haven't personally inspected and/or driven.
5) Never let yourself feel pressured into buying a car.
6) Don't ever leave a half-full coffee cup or bottle of Guinness on your workbench for more than three weeks. If you do, check it before you take a big swig. Particularly in hot weather.

Okay, those are the rules. Now the reality of what's actually in my workshop at this moment:

Exhibit 1 is a 1973 Lola 204 vintage Formula Ford. I bought the 204, fully disassembled, in a collection of rotting boxes that had been in a barn for about 10 years. I had to resist the impulse to load this car into my van using a Bobcat with a scoop shovel. Nearly four years later, project is almost done, but I'm still trying to figure out why the shift linkage hits the bellhousing and why there are three different sizes of banjo fittings for the brakes.

Exhibit 2 is a 1964 Lotus Elan, another barn find that hadn't turned a wheel in more than a decade. I've removed the fiberglass body from the badly rusted, disposable frame and stripped the interior. The entire instrument panel, complete with wiring harness, is resting in a cardboard box, over in the corner. This was an "exploratory" disassembly, and I don't plan on doing anything further until the Lola is done. The car is another future labor of love, waiting in the wings.

"At last," I muttered to myself, "after 57 years of abject defeat. And all I had to do was pay slightly too much for an old car in Texas."

Exhibit 3 is a heavy white coffee mug with a picture of the Alamo engraved on it, a souvenir of my last visit to this famous shrine during the Texas Hill Country Rally. It's sitting on my workbench, half full. I probably should have finished this coffee last December, when it was relatively fresh.

Anyway, you can see that I've broken nearly all of my own rules, namely 1, 2, 3 and 6.

Until this week, I hadn't figured out any logical way to break rules 4 and 5, but that was before I discovered eBay.

Yes.

I've been looking for an early '50s Cadillac, you see, partly because I really like old Cadillacs from this era, and partly because I need one for a special future road trip whose route and purpose are so secret even the CIA is stumped. And you know how hard they are to stump.

For better or worse, I mentioned this Cadillac quest to my friend Lew Terpstra. Lew is an administrator at our local technical college and a hard-core car buff who has fully embraced the Information Age and can play the Internet like Van Cliburn doing Rachmaninoff.

Tell Lew you need a door handle for a 1973 Avocado Green Frigidaire, and you'll hear the clatter and rustle of keys in the background—a sound not unlike squirrels on the roof—even as you speak. Seconds later, Lew will say, "There are six of them on eBay, and two more on RefrigeratorTrader.com." America's entrepreneurial currents are wired straight into his nervous system.

Actually, nearly all my friends are like this now, and I'm the only one who isn't. Makes you wonder if my somewhat insulting guidance counselor in high school wasn't right about all kinds of things.

Anyway, I told Lew I was looking for an old Caddy, and about 10 minutes later there were a dozen examples zapped onto my computer screen.

One car in Florida looked pretty good, so I decided to officially register for the first time as a bidder on eBay. Big moment. My buddy Pat Donnelly—who buys a lot of race-car and motorcycle parts over the Internet—walked me through the registration process, like an adult teaching a child to tie his shoes. It worked, and the next thing you know I was bidding like a house afire.

This Cadillac looked to be going really cheap, until about 30 seconds before the bidding ended, and then some other ghostly cyber-person started bidding against me at the speed of light in $100 increments. I lost the car in the last 3 seconds of bidding.

How could this person be so fast?

"You have to enter your highest final bid in a separate box," my computer-savvy friends explained to me the next day, "and then your bid will automatically be raised to that limit in whatever increments are being used—usually $100, with cars."

My eyes narrowed like a sneaky railroad baron in a Roy Rogers western. Aha!

The next day, Lew sent me another Caddy attachment to open, with a note that said, "This one deserves a look."

It was a sun-faded light green 1953 Fleetwood 4-door sedan from the hill country of Texas—with trunk-mounted air conditioning, no less. There were 22 hours left on the bidding. The top bid was low and the reserve had not been met, but there were 30 people bidding. The photos looked good—a nice, weather-beaten original with no visible rust or dents, so I called the owner.

He seemed like a nice guy and told me he'd bought the car from a man who'd inherited it from his aunt—the original owner. It had sat in a barn for 12 years, but he'd just done belts, hoses, tires, fluids and so on. It drove fine and everything worked but the radio and the right rear hydraulic window lift. His "Buy it Now" price was fairly high, so I returned to eBay and started bidding on the car.

This time, of course, I was a seasoned, combat-savvy bidder, so I knew exactly what I had to do.

I simply decided how much I would pay for the car if it were sitting in my neighbor's driveway and entered that amount as my final highest bid. It was a fair price, rather than a greed-driven dream of insolent theft, as my first attempt had been. It was also an amount that just met the seller's reserve price.

Lo and behold, at the last second before noon on Friday, my bid hop-scotched all the others and a message flashed on the screen that said "You, Peter Egan, are the Winner!"

"At last," I muttered to myself, "after 57 years of abject defeat. And all I had to do was pay slightly too much for an old car in Texas."

Or maybe it wasn't too much.

The next day, when I called the owner to get his mailing address for the check, he said his phone was ringing constantly and two other guys were bidding against each other on the Cadillac, in case I didn't want it.

Nevertheless, I got the car, and the owner called today to say that he received the check, and the transport company is coming this evening to load up the old Caddy. This week, after spending its entire life in dry, sunny, Kimball County, Texas, the poor car will be spirited away to the still-frozen north of Rock County, Wisconsin.

You will notice that this is a great deal of financial and physical activity to have set in motion with the impulsive clacking of a few computer keys. Sometimes I wake up in the night, breaking out in sudden cold sweats and wondering if I've lost my mind.

Why?

Well, I've finally broken rules 4 and 5.

After all these years, I've bought a car I haven't personally inspected or driven. Also, having just one day remaining on the auction imposed a level of pressure I've not experienced before in buying a car.

The whole thing is a leap of faith, based largely on a few computer screen images and a phone conversation with a man who gave the impression of being decent and honest. He probably is, and, with any luck at all, my instincts will be vindicated.

But then, George Bush looked right in Vladimir Putin's eye and thought he recognized a great soul. Perhaps the old KGB, non-democratic part of Putin's soul was hidden just off-camera, behind the far left corner of his retina.

In any case, the deed is done, and by the time I go to sleep tonight (or don't, again), a large enclosed car transporter will be hurtling through the Texas night with an 18-ft.-9-in. 4300-lb. surprise package strapped down in its cargo bay, headed for our northern home.

When it gets here, I'll either be the happiest guy this side of the Alamo or something less than the Winner my computer screen said I was.

August 2005

Rust Sometimes Sleeps

"How do you like your new creeper?" my neighbor Chris Beebe asked over the phone the other day.

This isn't a question you hear very often among the beautiful people at, say, the Cannes Film Festival or on the beaches of St. Tropez, but then we weren't at either of those places. We were in rural Wisconsin, living on a dead-end road near a rusted-out bridge where people fish for carp and bullheads.

Chris was curious about my creeper because he'd advised my wife Barbara to buy it for me as a Christmas present. It was quite a deluxe creeper, made of hard blue plastic, and it had a dropped pan so it slid around very close to the floor. It was slammed, chopped and channeled.

"I love that creeper," I told Chris, "but it's too low to the ground. Now I can't reach the bottom of the car any more, even with my arms out straight. I can only admire the oilpan and driveshaft from afar."

Chris chuckled wistfully. Being too far away from the underside of a car is one problem few home mechanics will ever have. Most of us go around with the tips of our noses burned by hot exhaust pipes and our eyes crossed from trying to examine drain plugs 2 inches from our bifocals. It's a close and claustrophobic world under most cars.

Especially if they are really big, heavy cars that make your floor jack groan and your heavy-duty jackstands look woefully frail and inadequate. I am speaking here of the 1953 Cadillac Fleetwood from Junction, Texas, that I recently bought on eBay.

Yes, the car I wrote about last month, bought sight unseen, arrived last Friday night in a 46-ft. Autobahn semi-trailer. Its 19 feet of sheer swanky glamour shared this huge trailer with only one other car, a C5 Z06 Corvette. The Cadillac fired right up with a wonderful mellow rumble from its dual exhausts, backed out of the trailer, and Barb and I took it for a drive.

It's a nice straight, unmolested car with faded, sun-baked original pale green paint. No dents, Bondo or accidents. The doors, hood and trunk all open and close like vault doors and all the body lines are straight.

It belonged, until four years ago, to a little old lady in Texas who died and left it to her nephew. He immediately sold it to the man I bought it from, a gentleman named Bobby Hurley, so I guess you could say I'm the third owner.

"If this car had been human, people would have been filing past its casket saying, 'My, but don't he look natural!'"

It had been in barn storage for 12 years, so Mr. Hurley did new brakes, belts, hoses, battery, plug wires, cap, tires and fluids. Now it runs and stops just fine. He also had the seats reupholstered, in correct original two-tone green.

Bad stuff? The radio and three of the hydraulic window lifts don't work, nor does the fuel gauge. This stuff is all fixable, however. Overall, it's a grand old car and I'm very pleased with the purchase and the missing element of tragedy in this transaction.

And a big part of that pleasure has been generated by that view from underneath—rolling about on my new, blue super-low creeper. I got under the car last night and just stayed there for about 20 minutes, gazing upward in the warm golden glow of my trouble light, floating in a state of euphoria, my brain humming like a battery charger, filling every cell of my being with pure, inexpressible bliss.

Why so happy?

Well, because there's no rust.

Here's a car that came down the assembly line when I was 5 years old. It was sold (according to the chromed logo on the trunk) by Alderson Cadillac in Lubbock, Texas, at a time when Buddy Holly was running around that very town as a 16-year-old. Meanwhile, in Washington, Ike was our brand-new President. This, folks, is a long time ago.

And yet there's no rust on the Cadillac.

Okay, there's a little rust. There are a few pits in the chrome, and the underside of the car has a patina of thin, brown surface oxidation, but it's solid, straight and structurally sound as a slab of granite.

I guess you have to have grown up in one of the Rust Belt states of the upper Midwest, where the winter roads are salted, to appreciate what that means. My whole car-crazy young adulthood was spent dealing with ugly cases of structural rust in virtually every car I owned.

My first TR-3, for instance, had a recent, sparkling paint job, but a closer look (just after I got it home) revealed serious problems. The floors—beneath all that glued-down carpeting—were made of galvanized barn siding and Pop rivets, and the taillights were screwed through carefully sculpted mounds of Bondo into wood blocks hidden in the trunk—with actual wood screws. It was a fairly accurate mockup of a former Triumph.

If this car had been human, people would have been filing past its casket saying, "My, but don't he look natural!"

My second and third restorations, another TR-3 and a Bugeye Sprite, were even worse. I spent hundreds of hours welding new sheet metal into rotten body and chassis sections. In the late '70s, I finally restored one last corroded Bugeye, hung up my oxy-acetylene torch and said, "That's it. I'm never touching a rusty car again as long as I live."

Right after that, in 1980, Barb and I moved to California—with our totally rusted-out 1968 Beetle, whose front fenders had to be riveted down with tin straps so they wouldn't flap in the wind and make us look like the Flying Nun. We immediately sold the Beetle for parts and bought a new Datsun 210. This car, with its California smog system, was a complete dog, but at least it wasn't rusty.

That VW Beetle was my last real flirtation with tetanus. Since then, all of my old cars have come out of the Southwest, or else they've languished in some Midwestern barn long enough to miss out on a couple of decades of salty roads. My 1963 Cadillac was like that; it spent 20 years sitting under a pile of junk in a Wisconsin garage, so the body rust was minimal. A rare AWOL from the salt wars.

The only other exception I've made to my No Rust rule was the 1964 Lotus Elan I bought out of a New Jersey barn last year. Elans have a fiberglass body on a steel backbone frame, and this frame is almost always rusted out, if the car has been driven for more than about 15 minutes on a humid day. But you can buy a brand-new, factory-built frame for $2800 and just bolt on the old fiberglass body.

So I didn't even blink when I removed the Elan's body and found this brown, iron-rich structure of crumbling potting soil where a steel frame used to be. No problem. I shall toss it into the metal scrap bin at our local dump with a great flourish of elbow movement and hearty, operatic laughter.

"Ha Ha! Come, Alfredo! Let us retire to my garage for a flagon of wine to toast my new powder-coated Lotus frame!"

So, except for that anomaly, my workshop has been happily free of rusty hulks for many years. The 1967 E-Type I restored a few years back came—like the Cadillac—out of Texas. Dallas, in this case. The Jag didn't have a spot of rust anywhere on its long, elegant body, and all the nuts and bolts came apart as if they'd just been tightened at the factory about an hour earlier. No cutting torches or hacksaws needed here.

You can find un-rusted cars in many parts of the country, of course, once you get out of the frozen tundra. I've seen nice clean cars in Georgia, the Carolinas, and even Florida, if they weren't too close to the ocean. I imagine all of the southern states are scattered with a pretty good selection of survivors, and California is another treasure trove of well-preserved old cars, providing they haven't been parked near the beach. Nightly fogs combined with salt air along the West Coast can be almost as damaging as Midwestern winters, so not all California cars are a sure bet. Inland is better.

Way inland, where the sidewinders roam. In fact, you might almost say that old cars and venomous wildlife flourish in exactly the same climates. If there isn't a scorpion or a desert diamondback nearby, you probably shouldn't buy the car.

There is just something right about cars that come out of the true Southwest—those bone-dry sections of Texas, New Mexico, Arizona, Nevada and inland Southern California. Desert cars. They may have faded paint and cracked plastic trim, but everything else is almost antiseptically good.

There's a special kind of dryness in the Southwest, a clean, perfect desiccation that's like a form of silence, as if time and all molecular activity have stood still in the hot sun. It's a preservative that works on old ranches, log cabins, mining towns, barbed wire, longhorn skulls, fence posts, farm machinery, cliff dwellings and even the mummies of the Anasazi. Everything old in the Southwest feels as though it's just waiting, gathering solar energy so it can resume where it left off.

Maybe that's why I spent so long on my creeper the other night, looking at the bottom of the Cadillac. I wasn't really examining anything in particular, I was just basking in the reflected light and warmth of the Great American Desert, soaking up a moment of suspended time, when I was 5 and Buddy Holly was taking guitar lessons in Lubbock. It's all written there, undisturbed, in the sheet metal.

September 2005

Vermin of the Dashboard

When I came into the house from the garage the other night, I looked at myself in the mirror and realized I had a raccoon mask of fine golden dust on my face, like a Wells Fargo stagecoach driver who'd just come skidding into Tombstone with an arrow through his hat.

Alas, I hadn't been doing anything so exciting as driving a stage in the Old West. Instead, I'd been upside down under the dash of my 1953 Cadillac Fleetwood trying to discover why the instrument panel seemed mute on such important topics as oil pressure, fuel level and water temperature. I also wanted to know why the wipers operated at a level of lethargy usually seen only in southern climates where rich people wear white suits.

What I found was an interesting mixture of what my friend Steve Kimball used to call "Okie repairs"—a bad mixture of friction tape, splicing and lamp cord done by some former owner with low standards of precision—and enough hardened mud to rebuild the Lost City of Abu Tabari, including the pool hall.

Up under the dash, you see, were dozens upon dozens of mud dauber nests. These little adobe-like formations, about the size and weight of hand grenades, were everywhere. They were stuck to all sides of the glovebox, all over the back of the huge Wonderbar radio, glommed to the top of the speedometer, and gathered like a village of mud huts on the underside of the upper dash. There was even a nest built around one of the wiper cables. When you turned the wipers on, it went back and forth like a mud wasp carnival ride.

They came off fairly easily. All you had to do was chip at the base of each nest with a long gasket scraper or screwdriver and the nest would crack and crumble and fall in your face, bouncing off your forehead with a light thud and showering you with dust. Very pleasant. In half an hour, I removed about 20 lb. of these nests and tossed them in a cardboard box on the floor next to the open car door. "If only I were still in fourth grade and could take these to Show and Tell," I mumbled to myself. "Mrs. Moe would be thrilled."

Luckily for me, there were no living wasps in any of these nests. Not even any dead ones. The little fluted tunnels in the mud seemed to have been vacated long ago.

". . . dozens of mud dauber nests, about the size and weight of hand grenades, were everywhere."

But this car must have been really something when it was first taken out of storage. The previous owner told me it sat in a barn in Texas for 12 years (but then he also told me all the instruments and gauges worked, so we have to take this for what it may be worth). In any case, it must have been swarming with wasps, and I imagine someone had to leave a bomb of insect spray in the car overnight to make it safe for human investigation and repair. It's hard to sell a car that stings the prospective buyer to death, as any good marketing director will tell you.

What amazes me is how deeply into the car these creatures would tunnel to build a home. They got into the car's interior, found the small air-conditioning vents mounted overhead in the headliner of the car and crawled down through the clear plastic ducts in the rear window into the fully-enclosed evaporator, which is located in the trunk on these old GM cars, and built dozens of nests in *there*. It must have taken them half an hour to get out when they needed a drink of water, or a little lunch.

At any rate, it's quite satisfying to get all this hidden mud out of the car, and, collectively, it represents an astounding amount of weight. When these nests are finally all gone, my Cadillac will go like the wind.

This is probably not a performance tip I need to pass along to the

Renault Formula 1 team, but it will certainly help the old Fleetwood.

Strangely, this car is my second "barn find" in two years—the other being a 1964 Lotus Elan I pulled out of a barn in rural New Jersey—and that car, too, was afflicted with its own pestilence.

No mud wasps in this case, just mice.

The Elan contained more milkweed fluff, chewed-up rags and other mouse-fluff material than any car I've ever seen. I spent about three days cleaning it out of the car and filled two full garbage cans. Like the mud wasps, the mice got absolutely everywhere. They filled every duct, door panel, shelf, hollow, depression and dark spot in the car. The place was a mouse hotel.

Oddly, I found no live mice anywhere in the car, and only a few ancient, desiccated skeletons here and there. My friend Chris Beebe—who helped me retrieve the car with an open trailer—suggested that they might have bailed out with tiny parachutes somewhere on the Pennsylvania turnpike when the temperature dropped below zero.

Miraculously, the mice that invaded the Lotus found nothing edible or gnawable in the car itself. The wiring harness, seats, cloth top and carpeting bore not a single tooth mark, and yet the car was packed solid with masticated cloth, rope, shop rags and various forms of rubber foam and synthetic material that had obviously come from some other part of the barn.

Maybe Colin Chapman, in his infinite engineering wisdom, had chosen wiring insulation and upholstery with anti-rodent properties. Or, more likely, he'd spec'd the material cost so low that even mice found it inedible. It could be a matter of difference in English and American tastes too. Maybe a Lucas wiring harness is the mouse equivalent of Vegemite, and it just doesn't taste that good to American mice. And Wilton wool tastes like undercooked white breakfast sausage. Who knows?

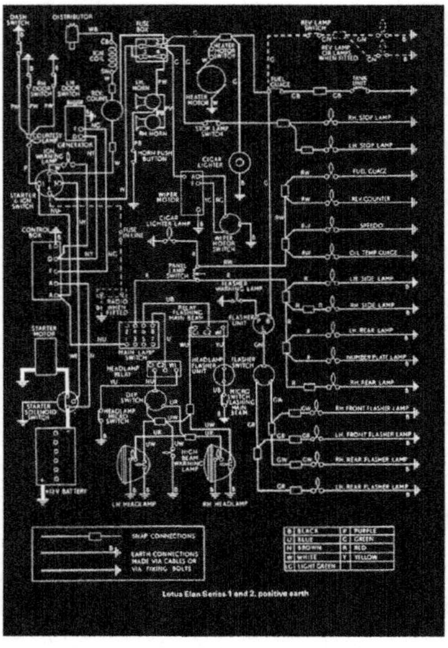

Lotus Elan Series 1 and 2, positive earth

Whatever kept those mice away, I wish they'd use it on American cars. My wife Barbara has a Jeep Grand Cherokee that develops an electrical short about every six months (usually in the winter) because mice build a nest under the air cleaner and eat the ignition wiring. They also get inside the air cleaner housing itself and occasionally build nests around the filter element.

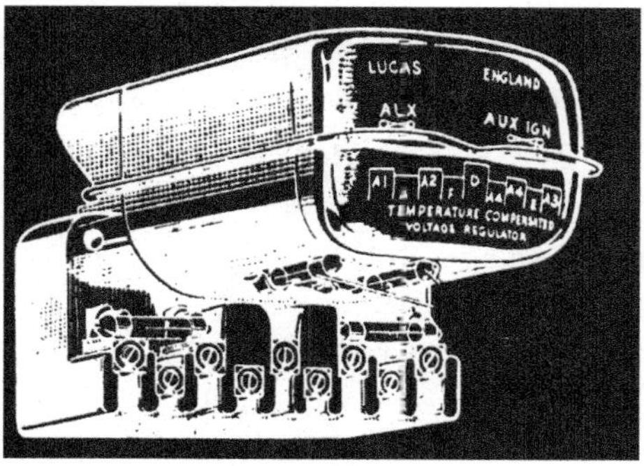

One can only imagine what it's like to take a long winter trip inside a Jeep air cleaner, or nested in the valley between two banks of a Chrysler 318 V-8. It's like stowing away in the undercarriage of a B-17 headed for the ball-bearing works at Schweinfurt. No wonder some mice seem to have that thousand-yard stare, or a variety of small facial tics.

Chris Beebe, who owns Foreign Car Specialists in nearby Madison, Wisconsin, tells me he gets quite a few cars towed in every winter with mouse damage. German cars, he says, are especially likely to have their wiring eaten by rodents, and he frequently sees New Beetles with one headlight on. Japanese wiring, he adds, never seems to be attacked by mice, but Honda Civics, in particular, have a series of sound-deadening chambers in the ducting that leads to the air cleaner, and these turn into little mouse apartments. "They like to live in one chamber and store food in another," he says, "and the third one is a restroom."

Other than wasps and mice, I can't think of too many other critters that attack our cars. Morgan owners always have to put up with termite and carpenter ant jokes, but I don't know of any genuine, documented cases of wood damage from these insects. I suppose it can happen, if a car sits long enough. As with old wooden boats, it's usually dry rot or the tin worm that gets them.

Some years ago, I bought an MGB from a desert community in California and found a black widow spider in one of the wheel wells while I was examining the brake lines. This made me very careful about sticking my hand into dark corners of the car without looking first. Black widows, of course, don't damage the car itself, but a dead owner might find that good news academic.

Luckily, I've not found any black widows—nor any sign of mice—

in the Cadillac. I mentioned this to Chris and he said, "That's because wasps attack and kill spiders, and I imagine they kept the mice out too."

In other words, the mud wasps may actually have acted as guardians of a sort, making my car uninhabitable to any other creatures, including humans. Which is probably all to the good. I'd rather clean those clumps of mud out of my car than the masses of damp, stinking mouse fuzz I found in the Lotus. They're easily removed and have no smell. Also, I'm not aware that wasp nests carry any major diseases, such as the deadly Hantavirus occasionally found in southwestern American rodent dwellings. So, compared with mouse droppings, nice dry Texas dust is a far more pleasant thing to have raining down on your face during a long night under the dashboard.

Try it and see if I'm not right.

Or you could just buy a car that didn't spend 12 years in a barn, as so many of our more sophisticated readers have chosen to do.

October 2005

The Fabulous '50s?

When it's raining cats and dogs in Italy and you're at the Villa d'Este *Concorso d'Eleganza* at Lake Como, there isn't too much to do except stand indoors at the pavilion and get wired on those tiny cups of espresso.

Espresso is to regular American coffee what rattlesnake venom is to a nice cup of hot cocoa, and it probably explains a lot about the energetic flamboyance of Italian design, not to mention the entire Renaissance south of the Alps and the movies of Federico Fellini.

Luckily for us visiting journalists, BMW (sponsors of this prestigious event) had thought of something else to do on a rainy day, an outlet for all that stored-up caffeinated energy. They'd organized an indoor symposium at the nearby Villa Erba, called "1950s' Design," with a panel of heavy hitters in the art business. So we all shuffled over there, shook off our umbrellas and sat down to listen, our sport coats reeking of wet wool like a sheep ranch.

Speaking that morning were the famous Carl Gustav Magnusson, director of design from Knoll, Inc. in New York; Konstantin Grcic, an industrial and furniture designer from Germany; BMW's own head of design, Adrian Van Hooydonk; Pierre Keller, a professor of art and design from Switzerland; Lorenzo Ramaciotti from Pininfarina; and Paul Warwick Thompson, who was invited because his name is easy to pronounce. Also, he's the head of the Smithsonian's Cooper-Hewitt National Design Museum in New York City. Quite a bunch.

A few cynics among the gathered press suggested that BMW had put this design conference together as an elaborate defense of the company's controversial styling trends, but it turned out to be just a lively discussion on the enduring influence of 1950s' ideas in our lives, dovetailing nicely with the 50th anniversary celebration of BMW's 507 sports car and the Isetta.

The '50s theme sounded interesting to me, not only because I'm so old I actually grew up in that era, but also because I was at that moment deeply immersed in (a) the restoration of a 1953 Cadillac Fleetwood and (b) a book called *The Fifties*, David Halberstam's superb 800-page history of the decade. It all fit together.

Halberstam, incidentally, says the 1950s are sometimes dismissed as an era of homogenized culture, complacency and social conformity, but points out it was actually a time of great artistic ferment, rebellion and vibrant hyperactivity, a release of pent-up energy and dreams suppressed by the Depression and World War II.

"*The new aristocracy won't have to be tied to computers, cellphones and electronic technology.*"

We might have had Levittown and crew cuts, but we also had Kerouac, Kazan, Elvis, James Dean, Brando, hot rods, beat poets, Charlie Parker, Chuck Berry, Martin Luther King Jr., Tennessee Williams, Hank Williams, Arthur Miller, Ayn Rand, *I Love Lucy* and *The Wild One*.

No wonder Fifties Chic is still with us.

For all that cultural weight, I must admit that I've never been obsessively nostalgic about the decade and am not one to roll a pack of Luckies in my T-shirt sleeve or hang out in chrome diners with carhops on roller skates.

Though that certainly sounds better than finishing this column. I'll be right back.

There, that was a nice drive, and a good smoke, too.

Anyway, I've always had mixed feelings about the art and styling trends of that era. And "mixed," as we know, is sometimes a code word for "queasy."

Take my parents' house.

My folks didn't concern themselves much with thematic unity in our home. They were modest people who had an eclectic mixture of wedding-present furniture left over from the 1940s and newer stuff from the 1950s they'd bought just because it was in stores.

The older designs I didn't mind much, but even as a very young kid I always thought that many of the 1950s' elements in our home were borderline hideous. We had some curtains with a pattern that seemed to be a mixture of hockey sticks and boomerangs, and a coffee table shaped like an amoeba. I was always afraid it was going to subdivide and we'd have two of them.

Our blond DuMont TV had fake wood grain on its metal cabinet, with big gold plastic knobs, and it sat on a slightly shaky chrome TV stand. It looked like a creature that might be packing a concealed ray gun. Our living room wallpaper had a pattern of giant hollyhocks and thick green vines. When I had mumps or measles, I'd lie on the sofa, look up from my Red Ryder comic book, focus on the wallpaper for a moment and say, "What on earth were they thinking?"

But not everything was unattractive.

"E-tech will become the work of the drone, and to be enslaved by these devices will be the mark of the under classes."

Parked out in front of our house (we had no garage) was my parents' 1956 Buick Special, which I thought then—and still think—is one of the handsomest cars of that or any other era. And ditto for my dad's well-used 1954 Studebaker station wagon, a Raymond Loewy design that still looks good today.

The '50s did produce some timeless objects.

And as our BMW design talk got underway, host Carl Magnusson projected an impressive photo montage of iconic products of the 1950s, which included a Royal typewriter, an IBM logo, Tupperware,

the Lockheed Super Constellation, a Rolodex, Bic pen (3 billion sold each year), Bauhaus cutlery, the Fender Stratocaster, Loewy Greyhound, BMW's Isetta and 507, a Braun hi-fi, Brigitte Bardot (great product!), the Lambretta scooter, the Saarinen tulip chair, Mies van der Rohe's Seagram's building, a Ferrari with body by Pinin Farina, the Honda 50 Cub, a Sony transistor radio, Nikon F camera and…the first Mini.

I enjoyed this overview a lot, as I had both an old Nikon F and a Rolodex sitting on my desk back home, a Stratocaster leaning on a guitar stand in the corner of my office, a Mini Cooper S inspired by the original 1959 Mini in my garage, and I've always been a huge fan of Brigitte Bardot. I also have a lot of Bic pens.

One element glaringly missing in this little slide show, of course, was the American automobile. GM in the 1950s was the most powerful and influential corporation in the world, and finned Cadillacs and '57 Chevys have now become almost a visual shorthand for the optimism and exuberance of the decade. But this was BMW's event, and I guess you can't blame them for leaving out my Fleetwood and including the 507. It's a beautiful car, too, even if it lacks road-hugging weight.

After the slide show, our speakers got up and gave their talks. I can't give the whole text here (carefully transcribed on the back of my airline ticket), but I was struck by a handful of key ideas and observations.

Lorenzo said, "We are already living in the future, but it's not the future we were predicting or imagining 50 years ago." Hooydonk then noted that when movie directors wish to show a scene is taking place in the future, even now they automatically include furniture from the '50s—those sterile white Saarinen tulip chairs and tables. "The future," he said, "is always our past."

We once imagined, Lorenzo added, that the future would bring more speed in all things—cars, airplanes, communications, etc.—but speed is bringing us diminishing returns. "Now there is often a feeling," he said, "that we don't have to do it any faster." Magnusson nodded and said, "The Concorde is gone. We are entering an age of consideration after an age of speed."

Thompson agreed and added, "To live wirelessly is the new luxury; to give up your Blackberry. The new aristocracy won't have to be tied to computers, cellphones and electronic technology. E-tech will become the work of the drone, and to be enslaved by these devices will be the mark of the under classes. Great hotels, such as the Villa d'Este here, will be inhabited by people who don't have to be connected. They will be left alone."

I all but stood up and cheered. As the last journalist on earth without a cellphone, I'd been ridiculed by PR people everywhere for not being reachable every second of my life. Now it turned out I was a member of a new aristocracy, rather than just another Luddite idiot. Thompson was my kind of futurist.

But another idea that hit home came from Magnusson in his closing remarks. He thought part of the reason for our current 1950s nostalgia in design is that the baby boomers were children then, and most children are protected from much of the struggle of daily life, so they remember those years fondly. It's more pleasant to reminisce over the Lone Ranger and your first Schwinn than it is to reflect on house payments and car insurance.

True enough. My father probably remembered the three transmission rebuilds we had on our Buick's Dynaflow (each on a separate family vacation) better than I do now. They didn't cost me a dime, and having our car towed to the nearest town was a big adventure. It's possible I liked this car better than he did.

But not everything was rosy for us kids, either. We had polio sweeping randomly through our neighborhood, and the newsreels at the Saturday Roy Rogers matinee showed H-bomb tests in the desert, with trees and houses blown away like tumbleweeds.

Also, we couldn't drive.

The past is all very well, but I'd rather be 57 and have a driver's license than be 10 again, without one. What I remember most about the 1950s was waiting.

November 2005

Running with the Big Dogs

So there I was, downshifting and giving it everything I had to move into the slipstream of Ferrari's Rubens Barrichello and Mark Webber of the Williams-BMW team. They must have heard me wheezing from the effort, because they picked up the pace almost immediately and casually disappeared into the distance, chatting amiably as they rode.

Yes, rode.

We were on bicycles, you see, rolling through the green hills of my home state, not far from the famous Trek factory in Waterloo, Wisconsin, suppliers of fine bicycles to Lance Armstrong and his Discovery Channel team.

Riding with us also were a couple of members of the Discovery team itself, Michael Creed and Tom Danielson. Needless to say, I couldn't keep up with those two guys, either.

Now, I don't normally attempt to ride bicycles with famously fit Formula 1 drivers and professional bike racers (or even the winners of local spelling bees, for that matter), but this was a special deal.

Someone at Trek had noticed that (a) quite a few F1 drivers use bicycles in their fitness training, and (b) several of these guys were more or less trapped in North America in the week between the Canadian and U.S. Grands Prix. Why not invite them down to the Trek factory, present them with some cool new high-tech bicycles, invite a few members of the press and some local F1 fans/bicycle aficionados and take a day ride into the Wisconsin countryside?

AMD, a software company that helps both Trek and F1 teams (Ferrari and Sauber Petronas) with wind-tunnel research, would cosponsor this little gathering. Good publicity and fun for all.

Mark Webber, Michael Creed and Tom Danielson

Fine, you say; nice idea. But why invite a guy like Egan?

He's 57 years old and has gained 12 pounds over the winter. His Spandex bike shorts and colorful jersey look like they were stolen from a circus monkey. He used to ride at least 100 miles a week, but this year a combination of business travel, hepatitis C and almost indescribable sloth has prevented him from even taking his exotic titanium Bianchi bicycle down from the garage ceiling, where it has been hanging all winter like a sleeping bat. The guy is a wreck.

I tried to tell that to my friends at Trek, but they told me to come along on the ride anyway.

Yes, despite my current low level of fitness (and owning an Italian bicycle), I do have friends at Trek. Interestingly, we all seem to have overlapping passions for bicycles, formula cars and high-performance motorcycles.

There appears to be a common gene that attracts people to the spare, reductionist element in all three of these sports, an appreciation for machines that combine a high level of technology with a conceptual purity of purpose.

For instance, my friend Jim Haraughty is an engineer at Trek who also road races motorcycles, and Trek press officer Zapata Espinoza is both an R&T and *Cycle World* reader. Another friend, Tom Schmock, is an avid bicyclist, Alfa enthusiast and Ducati rider who holds F1 big-screen TV breakfasts early on Sunday mornings in his Blue Moon Bar and Grill in Madison. Michael Schumacher, Rubens Barrichello, Christian Klien and Mark Webber all train on bicycles. Discovery Team's new star, Tom Danielson, used to race motocross....

The connections go on and on. It's all one big disease. You might call it an infatuation with things that go fast and don't weigh much.

Which doesn't exactly describe yours truly at this moment, but I finally took my bicycle down from the ceiling one hot, muggy morning last week, loaded it into my blue Ford Econoline van and headed for the Trek factory in Waterloo, which is only about 30 back-road miles from my home.

Rubens Barrichello

When I got there, a crowd of Trek engineers, bicycle magazine reporters, Team Discovery riders, AMD technicians and local F1/bicycle fans were all mingling in a large meeting room with Rubens Barrichello and Mark Webber. Christian Klien was supposed to be there, too, but had missed his airline connection from Montreal. Tom Schmock showed up with a gaggle of bicyclists from his F1 breakfast club.

I introduced myself to Barrichello, and congratulated him on a great race in Montreal, where he'd started at the rear of the field, with no qualifying time, and had still managed to finish 2nd.

"I had transmission trouble in qualifying," he told me.

"That was a lot of cars to pass in one race," I said.

He smiled and nodded. "It was very hot and my drinking water system didn't work, so I was pretty thirsty by the time I got to the podium." We chatted for a while and I found Barrichello very polite, articulate and down-to-earth. A Brazilian who now lives in Monaco, he speaks perfect English and has an alert yet low-key and friendly manner.

Same with Mark Webber, from Australia. A nice man to talk

with; like Rubens, he's smart and quick and exudes a calm confidence, but does so without a hint of self-importance.

After talking with these two drivers, it occurred to me that, regrettably, not nearly enough of their real personality and charm make it through the filter of TV sports coverage. Also, they're more distinctive-looking in person than they are on the screen. I wondered for a moment why that was, and then I realized they weren't wearing any baseball caps or sunglasses, which tend to make all humans look somewhat alike.

Note to F1: Lose the sponsor caps and fancy shades. The reason we can still tell Stirling Moss from Phil Hill is because they didn't wear this stuff.

Trek and AMD gave a technical presentation on their joint wind-tunnel research, Barrichello gave a short talk on the importance of heart rate and endurance in F1, as in bicycle racing, and then we all suited up for our afternoon bicycle ride into the country.

I was pretty worried about my own heart rate and endurance on this first ride of the year, but I needn't have been concerned. I was absolutely fine for the first 40 or 50 feet.

After that, however, things got bad.

I began to wheeze audibly (but certainly not "deafeningly," as some would later claim) and my leg muscles began to burn. It is said in bicycling that if you are out of breath, you're turning too low a gear, and if your legs burn, you're pushing too high a gear. I was doing both, simultaneously. There was no gear that worked.

And these guys could really ride. All of them.

Nevertheless, I gritted my teeth and somehow managed to hang on to the tail of the pack. Barely. But I wasn't really feeling all that chipper. I began to look around to see if any of my fellow riders had a gun to put me out of my misery, but bicycle racers usually don't pack heat.

Then a miracle happened.

A huge thunderstorm with violent lightning moved in from the west. Suddenly, it was dark as night. The Trek sweep wagon sped by and told us it was too dangerous to ride and instructed us to take a shortcut and head back to the ranch. So we returned to the factory in pouring rain, with lightning zapping the surrounding hills like something from *War of the Worlds*. When we got back to Trek, I had only 7.2 miles on my trip odometer.

Saved by the weather.

That evening, Trek and AMD threw a nice dinner party at a good restaurant in nearby Madison. Barrichello had flown off to Florida to meet his family and then take them on to Indianapolis for the F1 race, but Mark Webber spoke after dinner.

He said he grew up in a small Australian town and got into racing as a kid by riding BMX bicycles, and still uses a stationary bike to warm up for F1 races. He attended part of the Tour de France last year, and is a friend of Lance Armstrong's. Last February, he went bike riding with Lance at his Texas ranch, even jumping off a cliff into the water at the famous Dead Man's Hole, which is apparently a rite of passage among Lance's friends.

When asked how important fitness was in F1, he said "Very important. But it's still 20-percent driver and 80-percent car. In the Minardi I would often finish 3 laps down; in my last race with the Williams I was 5 seconds behind the winning car."

That weekend, I watched the U.S. Grand Prix at Indy to see how my new bicycle/F1 buddies (never mind that they didn't know I was there) would do in the race.

It was a complete debacle, of course.

Webber had to pull off after the first lap with all the other striking teams on Michelin tires, and Barrichello drove a fine race to finish 2nd, only to be booed by fans who were justifiably angry about the nearly empty field.

These drivers deserve better. Much better. It's always sad to see dedicated young men having their fitness and courage wasted by arrogant fools in suits, even though it's an old story.

Next weekend, I hope, will be an improvement. I see the Tour de France and the French GP are both on TV at the same time Sunday morning, so I'll be busy with the clicker, switching back and forth, keeping an eye on Lance, Rubens and Mark.

Maybe later in the day I'll try another bike ride to get in shape, and then work on my Lola. Inspiration, after all, is the only real point to professional sports.

Decmber 2005

The American Driver

When our motorcycle came over the rise and around the corner, I laughed out loud for a brief moment, then put my head down on the gas tank in the standard gesture of despair and defeat.

Ahead of us on this beautiful, curving, double-yellow roller coaster of a road along the rugged shores of Quebec were four slow-moving vehicles. They were, in this order: a large motorhome, a cop car and two big motorcycles. The motorcycles were pulling trailers.

I flipped up my face shield and turned to Barb, who was riding behind me. "Unbelievable!" I shouted over my shoulder. "Four of the most difficult-to-pass vehicles on Earth! And all in one group! It's like a bad cartoon!"

Barb patted me on the shoulder, by way of calming consolation. We could be stuck in this little train for many miles, unless we invented some excuse to stop and get off the bike. Maybe it was time to pull over at a scenic overlook. Get out a deck of cards, perhaps, or just finish medical school.

But wait! We didn't have to. This was Canada!

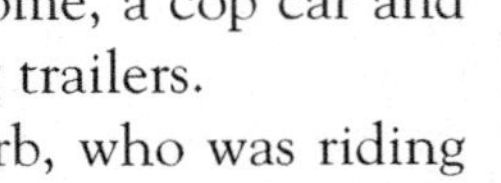

First, the police car turned off at the next little village. He was actually the only cop we'd seen in 2000 miles of riding through Ontario and Quebec.

Then the motorhome put on its turn signal and ran down the shoulder at a wide spot in the road, waving us past.

Without a cop and a motorhome, the motorcycles with trailers were easy to pass. They moved politely over to the right side of our lane and we glided by and waved. I noted that all four of these vehicles had Canadian plates.

These people were watching their mirrors! And they knew exactly what to do.

If this had happened in the U.S., we'd still be out there somewhere, following that same parade. Canadians, like most Europeans, not only have mirrors, but look at them once in a while. An art form almost entirely lost in the States.

Am I making a sweeping generalization here?

Perhaps, but recent personal experience tells me I'm not too far off the mark.

"Obliviousness, Sloth and Self-Righteousness. Yes, the Three Deadly Traffic Sins."

Barb and I just returned this Sunday to our home in Wisconsin, you see, from a 4000-mile motorcycle trip through eastern Canada to the Gaspé Peninsula on the Atlantic coast. We came home diagonally across New England and upstate New York. The last day, we took the Interstate home across Indiana and through Chicago to make time.

And we did make time, of course, until we hit the tollbooths around Chicago. These were backed up for miles in both directions. The last tollbooth in Illinois had southbound traffic stalled for at least five miles into Wisconsin.

Why the people of Illinois put up with this, I have no idea. Why would you pay your own highway department to bottleneck traffic, impede commerce, repel tourism, waste fuel, smog the air and make you late for vacation and work? Do the voting citizens consider this a valuable government service?

The whole highway system around Chicago is a national disgrace, but don't get me started. I might tell you what I really think.

Except for that short stretch of manufactured hell, however, it was a beautiful trip, with very few dull roads. And we rode all day long for 12 days, so we had plenty of time to contemplate the nature of the traffic around us. To compare and contrast, as my freshman History teacher used to say so chillingly in our semester exams.

So, herewith, a few observations:

In Canada, generally, the speed limits seem artificially low—typically, 90 km/h (close to our 55 mph) out on the open road—but everyone drives fast. Traffic on divided highways moves along at 80–90 mph, and hardly anyone with a fully functional vehicle is traveling at less than 70 mph on a two-lane road, unless it's very curvy or in a built-up area.

Yet these same "speeders" almost always slow down to a reasonable, safe speed in towns and villages. In other words, they drive at safe and prudent speeds for the conditions around them. When speed is harmless, they go fast; when it's risky, they slow down. They watch their mirrors, and on multiple-lane roads stay right except to pass. There's a maturity of judgment here—a sense of swiftness and dispatch without aggression—that seems totally at odds with the American driving experience. And yet there don't seem to be any cops—anywhere—to enforce this attitude. It's kind of like…Heaven.

Then you cross back into the U.S. and things change. The traffic gets slower, plodding drivers become more truculent (or maybe even recalcitrant), everyone drives in the left lane on the Interstate, and every other small town seems to have a cop running a radar trap. Suddenly, highway patrol cars appear in the flow of traffic.

To go from Canada (or England, France, Germany or Italy) into the U.S. is to feel exactly as if you've been demoted from adulthood and sent back to first grade, complete with hall monitors, teachers, lunch lines and slow-to-mature classmates who are still struggling with coat zippers and shoelace technology.

As an adult American driver, you feel you've been placed in a school desk adjusted too low for your knees.

Why is this?

There are, I believe, three basic forces at work here: Obliviousness, Sloth and Self-Righteousness. Yes, the Three Deadly Traffic Sins. The problem is, it's hard to know where one ends and the other begins.

For instance, you are following a motorhome on a winding road and the driver, who is averaging about 37 mph, has 43 cars backed up behind him, yet never uses a pull-out. Does this driver simply not see the other cars because he never checks the mirrors, or does he think 37 mph is plenty fast enough for anyone? Or is he simply worn out with the effort of pulling over every five miles on a 2000-mile journey around the U.S.? Maybe he's a kind of vampire in reverse, who sees only his own image in the mirror. Everything else is invisible.

Hard to tell, but in most parts of the civilized world (such as Canada) this guy can usually be counted on to pull over and make passing room at the first reasonable opportunity.

Not so in the U.S., where you might follow this driver in a long parade all the way from Colorado Springs to Cripple Creek, or until his refrigerator runs out of propane.

Interstates raise similar questions. When a car paces itself with a slow-moving semi (as we saw at least a dozen times on our recent vacation) and refuses to pass, does the driver not see all those cars in the mirror? Or do we have a self-appointed amateur cop on our hands, who thinks it's immoral to go faster than 64 mph?

My guess is the driver is simply too lazy to pass the truck, disengage cruise control or put on a turn signal and move over. This would require physical motion, as well as a small amount of judgment. It's just too much work. Besides, in the right lane you have to deal with merging traffic. Better to stay in the left lane all day, and let people sweep around on the right. If they can.

This sort of lethargy has led to an interesting condition on American I-roads: Our Interstates have now reversed themselves.

Yes, the right lane has become the fast lane, while traffic moves in a solid, slow train on the left. Barb and I breezed almost all the way across Indiana in the right lane of I-80/90, passing bumper-to-bumper traffic on our left. Occasionally we had to merge left and go around a slow car, only to observe a full mile or two of absolutely empty right lane. But no one would move over, probably for fear of having to make a passing decision sometime in the future. Or being cut off and losing a place to someone else. It's amazing. On a busy highway, nearly half the pavement goes unused.

I don't know what you do to change this. I suggested in a column a few years ago that we needed more emphasis on lane discipline in our high school driver's education courses. Students could be asked to repeat, at least three times a day, "Stay right except to pass," and that phrase could be emblazoned over the classroom door. I got several letters informing me that I was revealing my old age. "There are no driver's ed courses in most high schools any more," I was told. "It's a thing of the past."

That's too bad. How do you disseminate a cultural idea when

there's no mechanism to do so?

Private driver's schools? Peer pressure? Tradition? I don't know. Maybe these are all questions of natural courtesy that can't be taught.

My old friend and former R&T colleague Rich Homan used to say there were two kinds of people in the world, those who notice things and those who don't. (I seem to remember he ascribed this original observation to Lord Buckley, but I'm not sure). People who notice things, Rich said, will look to see if someone else is following them through a door and hold it open. Those who don't will let it swing shut in your face.

Maybe we need a national windshield sticker, one that reads backward, like an ambulance sign, so it can be read in the rearview mirror.

It would simply say, "notice things." Just like that, in small letters. No caps.

Type size is immaterial, of course, as our target audience won't see it anyway.

January 2006

Extinguishing the Midnight Oil

One of our secret pleasures here in the upper Midwest is the work of a singer and songwriter from Iowa City, Iowa, named Greg Brown. He comes through on tour once in a while—each overnight stop carefully chosen for its proximity to a good trout stream—and we usually see him when we can.

A few years ago we caught up with Brown in Fort Atkinson, Wisconsin, at a bar and restaurant called The Café Carp. He pulled up in what novelist James Lee Burke would call "a mid-'70s gas hog" and changed out of his hip boots on the street in front of the café. Extracting his guitar case from a trunk full of fishing gear, he strode into the place and sat down to play his soulful and poetic mix of tunes, wearing a slightly torn sleeveless T-shirt and an old felt hat, looking a bit wild and windblown, like someone who had just skydived into a briar patch.

The whole scene reminded me of the last Rolling Stones concert we went to, only without all the semis and security guards.

Brown has written a lot of great songs, but one of my favorites is called "Who Woulda Thunk It," and it's about getting older and losing your enthusiasm for excess and suffering. I won't print out the whole lyric sheet, but the essence of it might be contained in a couple of verses, here abridged (with apologies to Brown) for your convenience:

We used to say "I could walk all night,"
And we could and we did...
Now we say "I could walk all night,"
But it's not true
We can't walk all night, no,
Because we don't want to.

Another verse says,
We used to say "I could eat a horse,"
And we could and we did...
Now we say "I could eat a horse,"
It's not true
We can't eat a horse, no,
Because we don't want to.

The chorus after each verse is:
Hey, hey,
Hey, hey,
Who woulda thunk it?
Who woulda thunk it?

It's one of those songs that enters your repertoire of useful phrases and stays with you. Barb and I have used it a lot, in shorthand form, over the years. And, strangely enough, it's shown up three times in just the past month or so.

"I'd work until near exhaustion and then shower and go to bed, but lie there twitching like a high-voltage wire and thinking about what had to be done next."

On our recent road trip through Quebec and New England, I suggested to Barb that we stop for the night in a particularly bleak and seedy-looking motel along the highway, mainly because I was tired of driving and it was getting late. Also, I am cheap and hate to spend a lot of money on those lost portions of my life when I'm comatose.

Barb looked at this shabby little place from the parking lot and said, "We used to stay there, and we could stay there, but now we don't want to."

So I drove on another 25 miles and we found a place with better neon and the promise of fewer cigarette burns on the nightstand.

Right after that trip, the Greg Brown lyrics struck again while we were driving Up North for the weekend, and I suddenly swerved off the road to look at an MGB that was sitting by the roadside with a FOR SALE sign in the window.

The car wasn't rusty and the price was quite reasonable, but the interior, top, rubber trim and tires were all shot. The owner came out and opened the hood for me, and the engine compartment was an oily mess, with ratty wiring and a greenish-white radiator. It smelled like hot oil and coolant. He started the engine and it practically ran on almost three cylinders.

Not enough, really.

"Needs second gear," he said, "but otherwise the car works great."

I thanked him and returned to our waiting Mini.

"What do you think?" Barb asked as I accelerated down the road.

"Twenty years ago I would have dragged that car home and spent the next two years restoring it," I said, looking at her just a little sadly. "And now I could fix it up, but I don't want to."

She nodded and hid her sigh of relief behind a polite little cough, sort of like Doc Holliday during his time in Tombstone. She didn't want me to, either.

And, alas, this same expression of pre-emptive surrender and avoidance came yet a third time just a couple of weeks ago. The subject, once again, was an old car needing restoration.

The car, in this case, was my 1971 Lola T-204 vintage Formula Ford, which I've been restoring, slowly and methodically, for about three years. This car is kind of like that huge statue of Crazy Horse in the Black Hills—always nearing completion, but never quite done. My friends have begun giving me a fair amount of guff about this and are encouraging me to finish the project sometime before the next Ice Age brings an end to auto racing as we know it, or the Chinese burn off the last gallon of oil making plastic spatulas for Wal-Mart.

After I got the engine and Hewland gearbox successfully installed a few weeks ago, my friend Pat Donnelly looked at the project and said, "You know, if you would just get busy and maybe pull a few all-nighters, you could have this thing done in time to race the BRIC [Brian Redman International Challenge vintage races] at Elkhart Lake."

I glanced over at the Blues calendar on my garage wall, which showed a photo of Sonny Boy Williamson playing harmonica above the month of July.

"By golly, you're right," I said. "I *could* do it!"

So I started working feverishly on the car and ordering so many parts from so many sources that the UPS truck looked like a tuning fork going up and down our driveway. I didn't pull any all-nighters, but I did put in a lot of late-nighters, drinking iced coffee and caffeine-laced diet soft drinks into the wee hours of several mornings.

It was just like the old days when I was 23 and racing my Bugeye Sprite. I'd work until near exhaustion and then shower and go to bed, but lie there twitching like a high-voltage wire and thinking about what had to be done next, making mental lists and muttering to myself like a wino remembering the days when he used to be a big shot at Enron. I was on a crazy roll.

Things were moving.

But then, about a week and a half before the races, a few snags appeared.

First, the foam in my 30-year-old fuel cell turned to gasoline-flavored Jell-O when I added fuel to the tank, so I had to order a new one. And then I had starter trouble.

A Lucas starter came in an old box of parts with the car, but I refuse, from long and bitter experience, to put a Lucas starter in any race car. Especially a Lola 204, where the entire engine and transaxle have to be extracted to sneak a faulty starter past the frame tubes. So I ordered an expensive, reengineered, updated starter made from modern Japanese parts.

But when it arrived, it didn't fit. It hit the shift linkage and the frame.

So I sent it back and ordered a different, supposedly more compact, modern starter from a different source.

It arrived four days later and that night I took it out to the garage. Time and hours and minutes were running out.

I wrestled this starter down through the frame, and realized, to my growing horror, that this one wasn't going to fit, either. It was about a quarter inch too long. No go.

I sat down to catch my breath, glancing at the clock. Almost midnight. To have any hope of making the race, I'd have to put the old, cursed Lucas starter in. And to do that, I'd have to pull the engine and transaxle back out of the car, undoing all my recent work.

I'd have to burn the midnight oil. Pull an all-nighter. A bunch of them, actually.

Back when I was 23 and working on that H-Production Bugeye Sprite, I would have done it, too. All of us who raced did in those days.

Nighttime had no meaning, except as an extension of the day, a time when things got done. We'd start out bright and energetic in the evening and work until the sun came up, slowly losing our energy and zeal until we were wandering around like zombies, stumped by the simplest decisions. Go home feeling sick, dazed, dirty and stale, sleep for a couple of hours and then get up for the regular work day.

But that was then.

And now was now.

I got a Diet Coke out of my small Japanese workshop refrigerator, and sat back to look at my car. I started to pull the tab on the Coke can, but thought the better of it and put it back in the refrigerator. I got out a beer instead.

A dark, heavy *doppelbock*. Ambien of the gods.

"I used to work all night," I mumbled to myself, "and I could work all night. But now I don't want to."

I finished the beer, turned out the lights and went to bed. I'd be carless at Elkhart.

Hey, hey, who woulda thunk it?

February 2006

Insolent Chariots, Revisited

Good campfire material, this latest spike in fuel prices. Life always offers something worth discussing—wars, elections, hurricanes, etc.—but as we sat around the campfire a few weeks ago, the obvious topic was fuel mileage and the cost of gasoline. As well it should have been.

After all, we'd driven for five hours to get there and had just filled up our car. For about the cost of a new pair of Levi's, a medium anchovy pizza with hot banana peppers and a beer.

Yes, this is my value system. Now you know.

Anyway, there we were, camping in northern Wisconsin with our friends Jim and Patty Wargula.

Okay, "camping" is an over-rustification of reality. Jim and Patty have a large motorhome that Jim inherited from his folks. Every summer they take it up north to a campground in Door County and leave it parked for most of the season, using it like a weekend cabin. It has a shower, refrigerator, air conditioning, gas oven, microwave oven, reading lights, a DVD player, a TV and, yes, a kitchen sink. None of which is mentioned in my old Boy Scout Handbook, by the way, nor in the works of James Fenimore Cooper or John Muir. Camping ain't what it used to be.

But at least we had a traditional campfire, and we were all sitting around it, sipping on a nice cold beer from the refrigerator and discussing the high cost of gasoline, which had passed the $3/gallon mark that very weekend. It'll probably be drastically cheaper or more expensive by the time you read this, depending on the vagaries of war, hurricanes, supply, world demand and ordinary greed.

Jim wondered aloud why there wasn't more furor in Washington over the apparent gouging in fuel prices, and I said, "That's easy. Free-market types never met a windfall profit they didn't like, and conservationists are secretly delighted that SUV sales are falling. Nobody wants to rock the boat."

That particular weekend the price was about $3.04 for a gallon of mid-grade unleaded. I had just filled up our Mini Cooper S for $34, and calculated our mileage for the trip at 28.9 mpg.

Jim and Patty had driven up in their 1994 Cadillac Sedan De Ville (also recently bequeathed to them by Jim's folks). They, too, had filled up at the end of the drive, and Jim said that his on-board trip computer had reported an average mpg of 26.9 for his last tankful.

I took a sip of beer and stirred uneasily in my chair. Then I glanced over at the Cadillac.

"If a writer had to pick one year to make fun of Detroit iron, 1958 would probably be the year."

Was it possible that this huge car with its cavernous trunk, 4-door VIP sky-box seating and 4.9-liter V-8 was within 2 mpg of our lively but tiny Cooper S? To further muddy the equation, we'd been using the Cadillac all weekend as our run-to-dinner car and for our trips to the beach because the Mini was pretty cramped for four adults.

But then big engines, turning slowly and geared tall, often do very well on the highway. Our old 1988 Buick Park Avenue, which I sold to buy the Mini, got better fuel mileage (30–32 mpg) on the highway than the Cooper S, but averaged in the mid-20s in mixed driving.

And I was mollified somewhat when Jim told me his overall daily-driving average with the De Ville was only about 25 mpg. The Mini averaged 28 in daily driving. Also, friends of mine who own the standard, non-supercharged Mini report highway mileage of 35 mpg or more, and averages above 30. You have to pay something for that nice hit of wild-mouse acceleration the S model affords. For a sports car with four seats, it's reasonably efficient.

Still, you have to wonder what the Cadillac V-8 would do in a car that weighed no more than the Mini. Understeer, probably. But we don't make light cars any more (Lotus excepted). It's an almost forgotten engineering concept in this age of safety, remote door locks, 6-disc CD changers and computer-modulated cupholder deployment.

I have a 1953 Cadillac Fleetwood I'm restoring in my garage right now, and the first thing people say when they see it is, "Wow! Look at this huge car and those massive chrome bumpers! Can you imagine what that thing must weigh?"

"Yes," I always tell them, "it weighs 4330 lb. Or about 1000 lb. less than your new SUV."

We've come a long, way, Baby.

But back to the campfire.

Jim told me he might eventually trade their motorhome in on a traditional camp trailer. "One less vehicle to insure and maintain," he said. "Smaller to store over the winter—and easier to winterize. It's a whole other drivetrain I don't have to worry about. Also cheaper to move. This thing gets about 8–10 mpg on the highway." I could tell that last $150 fill-up had gotten his mental gears turning.

The next morning it was raining, so we did what we always do on rainy days in Door County, which is drive to Caxton's Books, a good used bookstore in nearby Ellison Bay.

While looking in the automotive section there, I found an interesting little volume called *The Insolent Chariots*. It was written in 1958 by an author named John Keats (probably no relation to the Romantic poet). I'd heard about this book, but never seen a copy. I carried it back to the motorhome and spent the afternoon reading.

The title of this book was taken from Lewis Mumford, the writer, city planner and chronicler of technological history, who famously used this phrase to describe the 1950s' American automobile. At a 1957 conference in The Hague, he spoke of "those fantastic and insolent chariots with which American motor car manufacturers now burden our streets and parking lots."

Keats's book, it turned out, was a humorous critique of those same "lower, longer, wider" American cars that ruled the road in the late '50s. If a writer had to pick one year to make fun of Detroit iron, 1958 would probably be the year. It was, as most automotive historians will agree, a year of over-the-top excess. And American consumers were, for the first time, really casting a jaundiced eye on the useless bulk and frivolity of Detroit's offerings and questioning annual design changes. Those little bug-shaped cars from Germany were selling like hotcakes.

There was rebellion in the air.

Author Keats, of course, has a great time shooting the bloated fish in this particular barrel. To illustrate what's gone wrong with the American automotive dream, he takes an imaginary family—Tim and Kim Vandervogel and their three children—on a cross-country vacation/camping trip in their new chrome-slathered, full-sized station wagon.

All the shortcomings of the wagon emerge on this trip. They have a flat and the spare is located under all the camping gear. Ride height is so low they scrape the exhaust system off on the crown of a logging road en route to their campsite. The thing handles like a pig, and its roofline is so low it feels like "an illuminated rolling cave."

What America needs, he tells us—and what Detroit doesn't build—is a more upright car, with a commanding view of the road, high road clearance, a larger gas tank, stiffer springs, a truck transmission and 4-wheel drive. "In short, it would be a machine able to go anywhere in any weather and it would resemble an Army weapons carrier, complete with water cans, shovels and picks strapped to its sides."

Well, Keats's dreams have come true.

Our campground was swarming with SUVs that fit his description of the ideal family vacation vehicle to a T—except maybe for the water cans, shovels and picks. But even those are options for the more rugged end of the sport-utility spectrum. America is finally saturated with the very wagons Keats envisioned.

In fact, we had one back home, sitting in our garage. It's Barb's 1995 Jeep Grand Cherokee, which she uses to get through the snow from our rural home to the local school, where she works as a physical therapist. She's taken good care of this Jeep, so it still looks like new, but it has 161,000 miles on the clock and it's starting to need more frequent repairs. Mechanically, it's getting tired. And on its last fill-up, it averaged 15.6 mpg.

So when we went up north for our woodsy vacation, we left the Jeep in the garage and took the Mini, which almost doubles that mileage.

Barb is still fond of the Jeep, but has decided her next trade will be on some kind of small wagon "that gets at least 35 mpg." She needs a wagon, rather than just a small car, because she volunteers for a dog rescue organization and hauls dogs and kennels around all the time.

So it seems we are moving on from Keats's description of the ideal

American car. Which, strangely enough, has become our new symbolic embodiment of the Insolent Chariot.

But even if the Jeep goes, I'll still have the category well covered with that 1953 Fleetwood in my garage, a car I love precisely because of its insolence. The beauty of the Cadillac, however, is that I don't need to drive it very often. The gasoline that goes into its tank is simply an occasional indulgence, like the gin in a good martini.

Unfortunately, it takes about 18 gallons of high-octane gin, so I may have to cut back slightly on my drinking.

March 2006

Crazy Mixed-Up Hot-Rod Speed Junkies on Wheels

It's become a tradition we don't even question any more; every summer, like clockwork, Barb and I go to Road America for the BRIC (Brian Redman International Challenge), our favorite vintage race. I can't imagine what would keep us away, unless I got my leg caught in a bear trap or was abducted by aliens with heads shaped like almonds and poor lower body development, as so often happens here in rural Wisconsin.

And another routine within that tradition is a lengthy cruise through the "vendor tents" at the track, where I generally abuse my Visa card by stocking up on armloads of car books and videos. They're usually about Formula 1 or sports cars, but this year something a little different caught my eye.

It was a DVD called *Born to be Wild.*

The cover touted "4 High Octane Movies" and showed a portrait of a young Jack Nicholson looming behind the image of a '34 Ford 5-window coupe with big rear slicks lighting up at a drag strip. The word "Wild" had speed lines coming off it, as you can well imagine it should.

What we had here were four late '50s and early '60s' hot-rod films. B-movies, in black and white. They were essentially outdoor movie fare left over from the heyday of that wonderful American institution, intended to provide background light for teenagers who were (a) making out like crazy, or (b) wondering if they should, and too deranged by that particular form of tension to pay much attention to the movie. How can you concentrate on dramatic content when someone has cute knees?

And why would you want to?

Anyway, these movies were made on low budgets with a specific audience in mind. And I would have been exactly that audience, had I been just a little older. But I didn't yet have a car or driver's license in that era, and it was pretty hard to persuade my dad to load the family in our Buick to go see *Hot Rod Girl*, *The Wild Ride*, *T-Bird Gang* or *The Choppers* at the local outdoor, so I didn't see any of these movies when they came out. I was too young.

But what is adulthood except a delayed end run around our parents' better judgment? So I naturally bought this DVD and dragged it right home. One evening, I made a big batch of popcorn and sat back to watch. Up first: *Hot Rod Girl*, a movie whose only known actor was Chuck Connors.

"No one seems to have a job. They just hang out, drive neat cars, play imaginary bongo drums."

Barb had her doubts about dedicating an entire evening to these films, but I said, convincingly, "They have *REAL HOT RODS* in them from the late '50s! How bad can they be?"

We soon found out.

Barb watched *Hot Rod Girl* for about three minutes and then went off into another room to water plants or wrap baby shower gifts, or whatever it is women do when you're watching the really good stuff on TV.

Okay, the dialog in this movie turned out to be a little slow, and the acting was pretty wooden.

And the plot was that predictable chain of events from many of the boys' car books I read in school: The town aristocracy wants to close down the drag strip, but a sympathetic cop (Chuck Connors, in this case) thinks it's a great way to keep would-be juvenile delinquents all in one place and channel their energy toward Good. Contrasts are drawn between real hot-rodders, who turn a mean wrench and understand their carburetors, and troublemakers who race on the street but are too morally lax to win at the drag strip. Gauntlets are thrown down; trouble ensues, tires screech and cars tip over. Square citizens are scandalized (also run off the road), the kid brother is killed, and the rodders finally learn their lesson. They grow up.

So *Hot Rod Girl* is not a really great movie, but it does open with period racing footage from the San Fernando Drag Strip, and you get to see the world's most interminable quarter-mile run between a white 1955 T-Bird and a white Jaguar XK-120. The T-Bird, unaccountably, wins (driven by our heroine, Lori Nelson), turning in a terminal speed of 98.94 mph. The movie also features a young Frank Gorshin, throwing in one of his classic "you dirty rat" James Cagney imitations.

The second movie, *T-Bird Gang*, is a crime movie with very few cars in it, except for another white Thunderbird, a '56 this time. The cops drive black 1957 Fords, and the opening warehouse heist is so *cinema noir* you can't tell if they're robbing a warehouse or looking for a flashlight.

The one familiar face in this film belongs to Ed Nelson, who has been in hundreds of TV shows and movies. He plays a crime ring leader who oversees burglaries from the cockpit of his T-Bird, apparently on the theory that no one will notice a sinister guy sitting in a white Thunderbird with the top down. Good cover. The best part of this movie is a superbly nervous jazz soundtrack by Shelly Manne and His Men. Cool in the extreme.

Our third feature, *The Choppers*, has a few more cars in it, and is livelier. Written and produced by a very young Arch Hall Jr. (who is also in it, as a gang member), it's about a group of young thieves who drive around in a poultry truck and strip cars that run out of gas along the highway in California. All of these cars seem to be 1959 or 1960 Buicks.

Maybe they got bad gas mileage.

The spoiled rich kid in their group drives a roadster said to be worth $5000, and the cops drive 1959 Chevys. When not stripping cars, the kids say things like, "I could go ape for you, Baby," and "Grease this palm, Daddy-O, peel off those cabbage leaves."

This movie is kind of fun, and filled with idiosyncratic performances, but its main appeal to car buffs will be seeing California roads with hardly any traffic on them. A lost planet.

By the time I got to the fourth movie, Barb had inexplicably gone to bed. Or maybe run off with a guy who had better taste. I wasn't sure. Nevertheless, I nuked myself a cup of leftover morning coffee and plugged in *The Wild Ride*.

Now we're talking.

This is actually a fairly good movie, cult-wise, and it stars a very young Jack Nicholson, in only his third film role. He plays Johnny Varron, the amoral and existential top dog in a hot-rod gang. In the first few minutes of the film, he runs a motorcycle cop off the road at night with his '57 Ford convertible, and then joins a beer party, unfazed.

Johnny is also a part-time race-car driver, and we get to see some pretty good dirt-track footage from a dusty Contra Costa Speedway. He runs his main competitor off the track, of course, to win. He doe not care about anyone. He's Top Man.

A rejected girlfriend comes up to Nicholson at a bar and says, anxiously, "Hi, Johnny…"

Nicholson looks down his nose at her, and his eyelids grow heavy. "Gimme a 10-spot," he snarls. She digs it out of her purse and he goes off to place a bet.

Brando couldn't have done it better.

Jack was only 23 when this movie was made, but the whole Nicholson persona is in place. A sidelight to this film is that Georgianna Carter, who plays the square "nice girl" in the movie, was Nicholson's girlfriend at the time, when they were both struggling young acting students.

A fun movie, with quite a few cars and fairly authentic-sounding beatnik dialog, unlike the others, whose scripts sound as if your high school guidance counselor wrote them.

But when the evening was over and the popcorn was gone, I couldn't help but notice that there had been quite a few similarities in all these movies.

First, they all have jazzy soundtrack music with lots of staccato bongo drums. The music is bleak and nervous. People are always snapping their fingers, and will dance at the drop of a hat, wherever they are. Even on the beach in broad daylight, which no guy I've ever known will do.

Second, they're all misunderstood kids, with indulgent but absentee parents whose neglect is behind all this juvenile delinquency. You never see the parents on camera. They are mythical beasts.

Third, the "teenagers" in these films appear to range in age from about 21 to 37. Also, no one seems to have a job. They just hang out, drive neat cars, play imaginary bongo drums on the countertop at the malt shop and say things like, "Hey, Baby, let's me and you cut a rug to this crazy jive!" while hopped-up sax music wails off the jukebox.

I mentioned this to my friend Tim Onosko (a connoisseur of B-movies and a documentary filmmaker himself), and he said, "Well, that was the California youth dream, wasn't it? Your wealthy parents bought you a car and all you did was drive around and have beach parties with your friends. No responsibilities, no job."

Yes indeed.

Unfortunately my own parents would never let me buy a car, even with my own money, and they made me work most evenings, weekends and summers in our family business. Also we had winter in Wisconsin and the beach was pretty much frozen.

No wonder I liked these movies. Or would have, if I could have seen them.

At the outdoor, you needed a car to see the car movies.

April 2006

Resuming the Elan

Anyone wandering into my garage the other night would have said I was crazy, but I prefer to think of myself as "differently saned."

I was, you see, standing at my workbench, wearing a pair of green coveralls, red rubber gloves and a black Arai full-face motorcycle helmet. Under the helmet was a dust mask of the Darth Vader pig-snout variety.

A pretty frightening picture, and part of the reason, I think, we have so few visitors from outer space landing on our rural Wisconsin property. They have their standards and can be freaked out just like anyone else.

In any case, space aliens tend to land in Western desert areas where property values and literacy rates are even lower, and they have a natural affinity for simple-hearted people and never show themselves to cynics like myself, even if we appear, superficially, to be adequately crazy to accept their weirdness.

Where was I? Ah yes, at the workbench.

So. Why all the odd-looking protection?

Well, I was disassembling a set of rusty Lotus Elan brake calipers and blowing them apart with compressed air. This is an ugly and dangerous job you shouldn't try at home, especially if you have nice wallpaper or tropical fish. I usually lock the caliper halves in my bench vise against a block of wood, cover the thing in rags and then feed compressed air in through a bleeder hole.

The corroded old piston awakens with a shudder, then slowly begins to ooze out of the caliper. When it finally reaches the end of its bore, it comes out like a cannon shot, usually releasing a fine toxic mist of ancient brake fluid into the air. Hence the dust mask and all the body protection.

I've never had a brake piston ricochet off the vise and hit me in the head, but there's always a last time. The forces unleashed here are fearsome.

This one exploded safely, however, and I found the caliper filled with a rusty sludge that looked like a mixture of wine silt and drain oil, only not as appetizing, beverage-wise. Unlike a fine Burgundy, brake fluid doesn't get better with age, and this car had sat motionless in a barn for 25 years.

The caliper bores were, as expected, badly pitted with rust. They'd have to be machined and re-sleeved. So rather than disassemble the other three calipers, I boxed them up for shipment to a place in Virginia where they restore your calipers for about twice the cost of your first car, if you're as old as I am.

Relieved of further combat duty for the night, I took off my helmet and dust mask and picked up a relatively recent cup of coffee. I checked the surface for brake fluid slicks, but there weren't any, so I sat back in my workshop beach chair (picture the turquoise and white on a 1956 Corvette) to ponder my latest garage project: the 1964 Lotus Elan.

Long-suffering readers will remember that I retrieved this car from a barn in New Jersey, in the dead of winter, almost exactly two years ago. It's a very early Series One Elan roadster (six taillights), bought new in 1964 at a dealership called Cox & Pulver in downtown Manhattan. The original buyer, Tom Cochran, met his future wife while dashing around the Big Apple in this car, and then gave it to her after they were married.

"She was driving a Sprite," he told me, "but it was so unreliable I made her sell it and gave her the Elan, so she could get somewhere without a breakdown."

Now there's a story you don't hear very often. When you need to provide a loved one with dependable everyday transportation, "Lotus" is not the first word that normally springs to mind. Nor does "Elan."

Even "V-2 Rocket" is higher on the list, as is "mining camp nitroglycerin truck."

Nevertheless, the Cochrans successfully drove this car all over the East for many years, commuting, driving to sports-car races and

taking ski trips to New England. Eventually the Lotus got too tired and rusty to drive. The fiberglass body, of course, didn't rust, but shock towers on the steel backbone chassis began to collapse. Tire tops pointed inward, like the legs on a spavined horse. If this were a horse, however, the owners would have shot it.

Instead, they called me, the Veterinarian from Lourdes.

I went to New Jersey with my friend Chris Beebe, loaded the Lotus on a trailer and dragged it home. It took me just a couple of days to clean the car, strip the interior and remove the body from the sub-frame to assess the severity of the project.

It was severe.

So I put the body back on the chassis to save space, threw a car cover over the whole thing and went back to work on my two other car projects. I was itching to attack the Elan immediately, but had to restrain myself until the others were done. Overlapping restorations are a dangerous business, and side effects may include headaches, mental disorientation and sudden hair loss. Ask your doctor if three disassembled cars are right for you.

Unfortunately, two is about my limit, so the Lotus has sat peacefully in the corner of my workshop, lurking beneath that car cover for the past two years, untouched.

Until last week, of course. That's when four things happened in quick succession:

1) I finished the mechanical work on my 1953 Cadillac Fleetwood and drove it over to my friend Denny Marklein's body shop for paint-

ing. This opened up a huge work space in my garage. So large, in fact, I was afraid the vacuum would suck in nearby barns, silo tops, farm tractors and livestock. The wind was deafening and sounded like a freight train.

2) I hauled my Lola 204 Formula Ford over to Steve Straavaldsen's race shop for some final fabrication (which I am too artless to do) and a complete suspension setup, leaving another big empty spot in the workshop.
3) After a warm, lovely autumn, winter dropped on us like a 100-lb. anvil. Snow and sub-zero weather arrived in tandem, and the sky went all Ingmar Bergman on us, dark and moody.
4) I took the car cover off the Lotus and looked at it again.

Oh, boy. That last one was a mistake.

I'd almost forgotten how rough this car was—and how beautiful at the same time. The yin and yang of the vintage Lotus experience. I quietly examined the Elan for quite a while, then went back to the house and spent a long, sleepless night, pacing the floor and looking through all my old Elan books and parts catalogs.

Decision time: Should I start working on this car, while the other two were temporarily gone, opening up a third front, as it were?

Or maybe sell the whole project to one of several friends who'd expressed an interest in it? It was a car that needed everything, after all, a huge undertaking. Once I started spending money on this thing, there'd be no turning back. Starting a car restoration or not starting one is the difference between sitting in an airplane with a neatly packed parachute or leaping out the door and shouting "Geronimo!"

With the airplane somewhere over Normandy. At night.

To complicate matters, our family financial advisor had recently warned me to stop frittering money away on old car and motorcycle projects, or I'd be working till I dropped dead. Put more into your retirement fund, he said. Much more.

Food for thought—and guilt. Maybe I should sell the Lotus and cut my losses. People will buy an inexpensive, entry-level dream, but not a costly, half-realized one.

Two more days went by.

Two more evenings with my workshop empty, the winter wind howling outside, and nothing to bead-blast, paint, disassemble or fix. I tried playing guitar for a while, but couldn't concentrate. Barb sensed my restlessness and suggested we drive to the bookstore, or go

to a movie.

But I didn't want to go to the bookstore or a movie. What I wanted was to resurrect an early Series One Lotus Elan, a lovely thing, and—to my mind—the epitome of sports-car design. And while it was true that I could sell the car and put the money into our retirement fund, who'd want to be retired without a project car like the Lotus?

It was a circular argument. I'd be saving up for a rainy day…filled with nothing but rain. Strangely, this is the basic question all of us in our late 50s seem to be facing right now. Do we squirrel away resources for the time when we are too old to do anything, or enjoy life while we can still remember which driveway is ours? Ideally a little of both, I guess. But as you watch your friends keel over at random, it's sometimes hard to find exactly the right balance.

So that evening I went out to the nearly empty workshop and put on my green coveralls. I then rolled out my big cherry picker and lifted the body off the Lotus with some canvas slings. I set the body aside on two sawhorses and began to remove the suspension from the subframe. Rusty old brake calipers first.

Taking the front left caliper over to the workbench, I donned my crash helmet, dust mask and red rubber gloves and went to work. The air compressor kicked in, and suddenly the quiet shop was filled with deafening noise.

Geronimo.

May 2006

Purr, Growl, Rattle and Hum

I was in the chiropractor's office yesterday getting my back straightened out, having accidentally given myself a half nelson while moving Lotus parts around the garage. Sitting crookedly in the waiting room, I was reading a *Wisconsin Natural Resources* magazine (it was that or *Good Housekeeping*—10 Meatloaf Recipes that Drive Men Wild).

In the nature magazine, an article said the great gray owl can hear a vole burrowing under a foot of snow, swoop down from a tree and pluck it out with perfect accuracy. There was a picture of an owl doing just that—an explosion of snow, a startled looking vole clamped in the owl's beak…violent stuff.

It's things like this that always make me thank God I wasn't born a vole—although being in Miss Podruch's third-grade class wasn't much better. She could hear whispering from 40 feet away with her back turned to the class while making the chalk fly and clatter across the blackboard. And she knew who it was, too.

We humans may not have the hearing acuity of owls or supernatural teachers, but I've always been amazed at the things we can hear and differentiate. Lying on the sofa in our living room (as is my custom), I can tell exactly what Barb is doing elsewhere in the house. The quiet thud of an orange juice carton against a countertop is quite different from that of a milk jug, and, from two rooms away, you can identify the sound of a cap being taken off a fountain pen, the measured, even breath of someone writing a check.

Yes, even to a guy who spent years standing in front of a full Marshall stack playing "Serves you Right to Suffer," these signature sounds are instantly identifiable.

And cars, of course, have their own signature sounds as well.

As I was reminded on a recent press trip to England, visiting the Jaguar factory to gather some technical background on the new XK.

Part of the visit was a trip to Jaguar's sound lab, where an engineer named Gary Dunne gave us a presentation on car sounds and what they mean to the human ear. It's Dunne's job, essentially, to make sure the Jaguar purrs and growls in exactly the right way.

He explained that a car's sound imparts three different values: quality, information and emotion. "Emotion," he said, "is an essential part of what Jaguar is."

We put on headphones and listened to recordings of different high-end performance cars, as heard from both roadside and the passenger compartment. Jaguar actually has a selection of different "artificial" heads ("named after former employees and members of the Simpson family—Bart, Homer and so on") that have microphones in their "ears" to accurately record sounds.

"We can't rely on single human impressions alone," he explained, "because we all hear differently. Even how our ears and shoulders are shaped affects the way we hear the sound spectrum...how long our necks are."

I nodded sagely. This explained why so many football linemen miss the subtleties of baroque chamber music.

Through the headphones we listened to sounds of those various GT cars. Interestingly, the civilized, ultra-quiet end of the spectrum was represented by a Lexus 430, a subdued, honey-like euphonious thrum. "A very high-quality sound," Dunne explained, "but quieter than we like at Jaguar."

At the other end was the 4.2-liter Maserati—hyper, high-pitched and snarling. "Wonderful sound," Dunne noted, "exciting and very Italian, but a little louder and more aggressive than our customers expect."

We then listened to the just-retired XK8, which had a very warm and pleasing growl. Dunne explained that this sound was just above the middle line between the Lexus and

Maserati extremes, tilted slightly toward the civilized Lexus end. "Our new XK is sportier and more aggressive than the old car, so we have now moved the sound closer toward the Maserati, just below the center line of the spectrum." He played us a tape, and it was more aggressive. Slightly less purr and a little more growl. Also a little more high-end rip. Fascinating.

But how do you "move the sound"?

First, Jaguar had used the primary thing I would think of—exhaust tuning tricks. While I usually let the exhaust system rust off or accidentally stab the muffler with a jack handle, Jaguar had been more scientific.

They'd separated the mixing of the two sides in the forward resonator and then modified the transverse rear muffler with a blow-off valve to open up the main chamber under the pressure of hard acceleration.

After the exhaust work came a less obvious refinement. They ran a hollow tube from the engine's intake plenum back into the cockpit to a small plastic speaker-like device that amplified intake noise under hard acceleration. This device added an exciting, banshee-like intake howl as an overtone to the deep and mellow growl of the exhaust.

"My 1964 Lotus Elan doesn't even have one of those speaker tubes," I joked with Mr. Dunne. "What was Colin Chapman thinking?"

To his credit, he laughed good-naturedly and didn't have me killed.

Another sound-tuning trick in the new XK, he said, was the addition of a small tuning-fork-like yoke attached to the rear transmission mount. This (as I understand it) cancels out some annoying driveline noise and amplifies the good stuff.

There were, of course, many more tricks to remove tire hum, chassis resonance, etc., but I don't have the time or space to go into them all here, even if I could read my own notes. The important thing is, Jaguar takes the sound of its cars very, very seriously.

I stumbled out of the sound lab duly impressed. In every road test we try to convey the nature of a car's exhaust note, but I'd never really considered what a huge effort is made to project quality through sound. It seems like a secondary attribute of the car, yet a sports or GT car with a disappointing exhaust note is, well…disappointing.

On the plane home I got thinking about a passage I'd read in Geoffrey Healey's memoirs about the development of the early Austin-Healey sports cars. He said that he and his father, Donald, had spent a long weekend at the factory garage, installing different combinations of mufflers, resonators and exhaust pipes on a Healey 100 prototype, trying to get the sound just right. They finally found the magical combination and his father approved it for production.

How different an approach this was. Very seat-of-the pants (appropriate for a former World War I aviator like Donald Healey). Listen to what you like and then bolt it on there. No fake heads, microphones, computers or sound-wave printouts.

On the other hand…the Healey 100-4 exhaust note probably could have been improved. I spent about 2500 miles touring in one with my buddy Chris Beebe a few years back and I found the sound pleasant, but a bit droning and flat on the highway, missing some sonorous element.

Chris, who still owns the car, agrees. We had lunch the other day and he said, "The Healey that really got it right was the 3000. Beautiful sound, one of the best ever. But the 100-4 isn't quite in the same league."

This naturally got us discussing our all-time favorite exhaust notes. I thought it over and said that mine would have to be my first TR-3. It had a simple muffler and resonator on a single pipe, but it put out a resonant, throaty growl unlike anything I've owned since. It hummed deeply on the highway, with a crisp edge produced by a touch of the throttle. It was like listening to an old Vox AC-30 amp with just a touch of distortion; very rich and bluesy.

Chris said his favorite was always the Jaguar Six, like the one in my old E-Type. "In the summer, I could sometimes hear yours start up early in the morning [Chris lives across the river from us], and when you backed out of the garage the sound was perfect. It's a little quiet when you're in the car, but it sounds great from a distance."

I nodded and said the E-Type had always been a little too quiet for me, but I sensed the exhaust system was muting the most beautiful engine noise in the world, smooth and harmonic. It sounded like something important was happening at the other end of those chromed stingers. The car had a beautiful purr, but needed more growl.

We talked about V-8s and agreed their sound quality, like that of inline-4s, was all over the map. One of the sweetest sounds in the world, to my ear, is a flathead Ford V-8 with glass-packs, while other V-8s I've owned merely sounded flatulent and wasteful, with that infamous "drink-a-gallon" idle. My 1953 Cadillac 331 actually sounds pretty good—subdued and growly, like a car sound effect from an old radio show.

This discussion inevitably came around to my Lotus Elan, now under restoration. Chris, who has two Elans, told me his Plus-2 sounds great, but said his standard coupe's stock exhaust system was a little disappointing. "We'll have to work on yours when the time comes," he added, "make some modifications or find something better."

Indeed we would. The Lotus was a light and pure sports car, I reflected, but it was too bad Colin Chapman didn't have Bart and Homer to work with back in the early 1960s. The sound of the new XK—to a person with my ears, neck and shoulders—is just about perfect.

June 2006

A Guide to Collectible Sleepers of the Lowest Rank

Just yesterday, one of my favorite classic-car magazines arrived in the mail, containing a list of "10 future collector cars you should buy now." Sitting back with my morning coffee and looking over the list, I was appalled to find it contained both an AMC Matador and a Chevrolet Chevette Diesel.

Well, not appalled. There's an element of offbeat good fun in stumbling across cars like these in your neighbor's backyard, but I can't see anyone actually "collecting" them on purpose. Still, I suppose this list was inevitable.

After all, we live in an age where car auctioneers are wringing seven figures out of old muscle cars that—let's face it—are, for the most part, indifferently constructed American sedans with big engines that allow them to spin sideways and smack fire hydrants, while delivering fuel mileage that makes a Hummer look like an Earth First plot against Exxon.

Not that I don't like these cars myself.

I, personally, would be willing to spend about $16,000 on, say, a nice 1968 Hemi-powered Dodge Charger, but when auction prices shoot past $100,000 my attention wanders and I start thinking about the 1956 World Series or wondering if I left the coffee pot plugged in before we sold our last home. In 1990.

An MGA for $32,000? These are lovely cars, granted, and I've always liked them, but isn't that rather a lot for a car that my roommate in college bought for $650 because he couldn't afford a used Triumph motorcycle?

Anyway, you can see my point. As auction prices on genuine "blue chip" classics continue to climb, those of us without Ferrari GTB or Cobra money—indeed, without even Pinto money—will begin to look farther and farther downstream, as the crossbar is continually lowered until it's positively subterranean and only those with excavating equipment need apply.

Luckily for the readers of R&T, we have someone on the staff who is highly qualified to get the jump on the market and compile a sleeper list for the sad day when all those Matadors and Chevette Diesels have been snapped up, and that would be me.

Yes, as one who spent 10 years as a professional foreign car mechanic during the 1970s (a truly terrible era of car design)—and one who spends way too much money restoring the wrong cars for all sorts of shaky reasons—I believe I am uniquely positioned to reflect light on this short and badly soiled end of the automotive stick. As is my friend and former employer at Foreign Car Specialists, Chris Beebe, who owns about 50 really odd cars and helped me compile this shopping guide.

So here's a list of Next Wave sleeper collector cars that hardly anyone has ever sought. Until now. Remember, you saw it here first.

1. **The Austin Marina.** This car looked good on paper, as it was a simple, basic sedan with an MGB short block. Unfortunately, the head used a single Stromberg carb with an overly complex electric choke and smog plumbing and a distributor with severe built-in retard. These cars simply could not be made to run right, nor to produce anything recognizable as horsepower.

Also, the single-rail gear selector fell apart and dropped into the transmission, and rear axle bearings failed. These very features, refined to the point of total debilitation, turned up later in the Triumph TR-7. I haven't seen a Marina anywhere in 30 years, so now must be the time to snap one up, before values skyrocket into the stratosphere, or even the biosphere.

2. **The Sabra**. This was a sports car made in Israel, with a fiberglass body and (Chris recalls) a Vauxhall Alexandria engine, named after a city whose library burned. The steering wheel was sawed in half to make the car feel more like a Beech Bonanza, and the raw tubular edges had rubber "chair feet" to protect the owner. I've

seen and driven only one Sabra, and it was admittedly quite worn out, but the car almost defined the word "loose." Everything rattled, moved and shifted around. If you can find a Sabra, run, don't just shuffle morosely, to the bank.

3. **Kaiser-Frazer.** What do you do when tank production stops after World War II? Make more tanks! My dad bought a maroon 4-door Frazer (which is the first car I can remember) shortly after I was born in 1948, and Jay Leno has one that looks exactly like it, donated to his collection by an admirer who wanted it to have a good home. Maybe it's even our old car. My own nostalgia for one—and Jay's tacit approval—are bound to make these things almost unobtainable in the coming several decades or more.

4. **Austin America**. I hate to pick on Austin here, but these cars—intended as a slightly larger and more modern version of the original Mini—had all the problems (and then some) of a Mini, but without the good looks. Think about that. Our local Pizza Pit bought seven of them with automatic transmissions, and they all disintegrated in exactly six months. Yours could, too, with plenty of TLC. To avoid paying too much, look for one with a faded "Pizza Pit" on the door.

5. **Rover 3500**. My friend George Allez bought one of these and was then somewhat distressed when an identical car, painted gray, kept turning up as a staff car for the East German secret police on the TV series *Mission*

Impossible. His car spent most of its time in the repair shop, and was then sold for a fraction of its purchase price. It's only a matter of time before Tom Cruise uses one of these in a *Mission Impossible* flashback and values go right through the basement ceiling.

> ***"Remember, ugly stuff always comes around again, while beauty seems impossible to recapture."***

6. **Triumph TR-7.** British Leyland made an all-too-common mistake here, thinking that "ugly" and "modern" were exactly the same concept, a misapprehension that has also haunted public architecture since the 1950s, when everyone read Ayn Rand. The doorstop styling might have been forgiven if the car had been screwed together better and differentials hadn't failed at low mileage. They also shared the Marina plague of transmission and axle bearing ills. Still, they handled okay, so these cars may be out of our bottom-feeder price level. What I do in this case is find a fatally rusty example and laugh all the way to the bank, if I get that far. Look for a car where the left front tire rubs on your clutch foot.

7. **Pontiac Aztek.** There's still time. There will always be time. Well, maybe not. It's possible a new-generation Dana Carvey and Mike Myers are being born right now and will do for the Aztek what they did for the Pacer in *Wayne's World*. Remember, ugly stuff always comes around again, while beauty seems impossible to recapture.

8. **Datsun 210.** These cars were pretty good in most of the coun-

try—simple, serviceable and reasonably lively. The California smogged version of 1980, however, was slow almost beyond belief. Barb and I unsuspectingly bought one as a new car when we moved to the Golden State. Pulling out onto the highway, you'd shift the Datsun up into 3rd gear and it would actually go slower. Depressing the throttle had exactly the same effect as dangling a donkey in front of a carrot. Anyway, the sleeper model of this car is the California version. Don't be fooled into buying the high-performance 49-state job, which makes literally dozens of horsepower and is way out of our price range here.

9. **Fiat 850 Spider**. I know a lot of people who liked these cars and claim that Fiat always gave you a lot of value for the money. Well, so do chorus girls, but they don't rust out. I worked on these cars as a mechanic and have not forgotten, even though I can't remember where I put my glasses or who directed *La Strada*. The basic driveline was pretty stout, but there were chronic problems with kingpins and axles. And rust. I can't help feeling they belong in any comprehensive list of sleepers a person such as myself should be able to afford, in a just world where no individual is discriminated against just because he has poor financial skills.

10. **A Renault anything.** Take your pick, from Dauphine to Alliance. "These cars seemed to work okay in France," Chris told me, "but something happened to the metallurgy when they crossed the Atlantic. Also the Great Lakes…and the Wisconsin River...God help you if you lived west of the Mississippi." Chris bought a new LeCar on a cold winter day and it wouldn't start the next morning. When he attempted to open the hood to spray some starter fluid into the air cleaner, the hood release, hood handle and wing nut for the air cleaner all broke

off in his hand. Other Renaults? "The Dauphine had a plastic reverse gear that always failed," he said, "and you couldn't jack up the Caravelle without all the doors closed or you'd twist the chassis." Most of these cars have disappeared now, but the wise shopper might still be able to find an Alliance rotting behind an old building somewhere, just waiting to accrue unanticipated future collector value beyond anyone's wildest dreams.

There are many, many more I could mention, but I want to do some scouting and see if I can find cheap examples before I start a stampede of savvy collectors like myself. Good luck.

Section II
THE FEATURES

August 2004

9th Annual Amelia Island Concours d'Elegance

Offering sanctuary to great cars and stars

Years ago, my wife Barbara and I took six weeks of accumulated vacation and flew our antique Piper Cub from California all around the U.S. On July 12, 1987 (so sayeth my log book), we found ourselves flying low and slow over the white sand beaches and river estuaries of the Southeastern coast. Racing an evil-looking offshore afternoon thunderstorm, we attempted to land on Jekyll Island, off the coast of Georgia.

On final approach, a gigantic lightning bolt/fireball hit the end of the runway and a mighty wind came up that carried us sideways across the channel to the mainland in what seemed like mere seconds.

Raindrops the size of pigeon's eggs began pelting our windscreen. It seemed like a good time to turn around. We raced the storm south and landed safely at Fernandina Beach airport, just across the Florida border, on an island called Amelia.

There, a fellow pilot at the airport befriended us and gave us a driving tour of the sleepy little resort community, with its shaded streets and trees hung with Spanish moss. "I love this place," he said. "It's one of the undiscovered treasures of the Atlantic coast."

We told him we needed to find a hotel for the night, and he said, "Oh, I own a couple of condos on the beach that aren't occupied now. You can just stay in one of those." And we did, listening at night to the waves rushing up the sand and the ocean breeze rustling the trees.

Seventeen years later, Amelia Island isn't quite as undiscovered as it was—big hotels and more condos now rise from the shoreline—but it's still a sanctuary of sorts and welcoming place of great charm.

Above, John Groendyke's immaculate 1935 Auburn Speedster Model 8-851; Below, the Indy-winning 1939 Maserati Boyle Special, normally on display at the Indianapolis Motor Museum.

Especially for car buffs, now that R&T contributor and Jacksonville resident Bill Warner has chosen to hold a concours on the green lawns of the golf course adjacent to the Ritz-Carlton Hotel every March. It's become a winter getaway for those of us in the frozen North and a great jump-start to the year's car madness.

Last year, heavy rains drove the car show into tents, parking ramps and the hotel ballroom, but this year Bill bought $23,000 worth of rain insurance and thereby guaranteed perfect weather. It was glorious. And the sun shone on an astounding array of about 250 of the most beautiful and significant cars on earth—sports cars, race cars, hot rods, 1950s' sedans and grand old touring cars with brass radiators. Something for everyone—and spectators were said to be backed up all the way across the Highway A1A bridge waiting to get in (proceeds going to Community Hospice). On Saturday, the RM Classic Car auction also drew a large crowd.

Bobby Unser was the Guest of Honor this year, and a collection of his Indy and Pikes Peak-winning race cars was displayed on the lawn. Unser gave a typically colorful and uninhibited talk about his early racing career at the black-tie dinner sponsored by Mercedes-Benz, telling how he bought "a little old $3500 airplane" to fly back and forth between Albuquerque and California. He wore the airplane out, he said, just trying to get his name known on the West Coast.

The audience was studded with racing stars—Johnny Rutherford, Bobby Rahal, Donnie Allison and Marvin Panch, to name a few—along with dozens of famous photographers, designers and automotive artists. Bill Warner introduced about 50 of these luminaries, with no notes and without missing a name.

Bobby Unser stepped up to the microphone and said, "How does he do that? I can't even remember Rutherford's name half the time—especially after a couple of drinks. And I used to race against him."

On Sunday it was the enviable job of R&T contributor and Chief

Judge Tom Cotter and me to pick "The Car We'd Most Like to Drive Home" for the R&T steering-wheel trophy. From an embarrassment of riches, we finally chose a shimmering green 1960 Aston Martin DB4 GT Zagato belonging to David and Ginny Sydorick, of Beverly Hills, California.

Fortunately, I didn't actually drive this car home, because when I flew back to Wisconsin it was 21 degrees and snowing, with dark Siberian overtones. Maybe next year I'll pick something with all-wheel drive. Or just stay on Amelia Island until spring, with the waves rushing up the sand and the ocean breeze rustling the trees. After 17 years, it's still a great refuge.

April 2005

Zamboni 500

At last, a sport utility vehicle that lives up to its name

We at R&T don't normally test vehicles that lay down their own pavement, but we made an exception this time because we are the only car magazine in Southern California whose Managing Editor has received six stitches after being hit in the head by a puck while serving coffee as a hockey mom.

Also, our Associate Art Director broke his arm last year playing ice hockey, and our Production Editor is a hockey groupie whose interest in the young men who play this sport borders on stalking. Luckily she's a woman, so the restraining order has been enforced with no great zeal, much to the relief of both players and staff.

In other words, this office is a seething cauldron of hockey enthusiasm and mayhem, so our long-suffering Editor-in-Chief finally caved in and agreed to let us test this unique machine, known as the Zamboni Ice Resurfacer.

Unfortunately, he also decided to fly Yours Truly in from

This sleek, black monster is a delight to operate with its myriad levers and necker-knob steering wheel.

Wisconsin in the middle of winter to write the test, apparently working on the theory that people who live in the upper Midwest have a natural affinity for ice. So I left Wisconsin on a morning when it was 2 degrees below zero, flew into sunny California and drove straight to the Long Beach Arena, which has the cheerful indoor climate of a Winnipeg meat locker, despite being less than 100 yards from the beach.

Most of our readers probably don't think of California as a hotbed of hockey enthusiasm, but it is. The Long Beach Ice Dogs are a professional feeder team for the Montreal Canadians (when they aren't on strike), and they play regular games with teams in San Diego, Fresno and Bakersfield. Yes, there's a rink in Bakersfield. That's probably why there are so many Buck Owens hockey songs.

We gathered outside the Long Beach Arena on a Friday morning and shook hands with John Debilzan, a professional Zamboni driver who's been in the business for 12 years. He opened the garage doors and gave us a tour of the big black machine.

It's a purposeful-looking thing, though its styling is certainly controversial. One of our staffers commented that it looked "beautiful in a utilitarian way, reminiscent of an early Land Rover," while others thought it looked like a "dumpster with a steering wheel." Our resident BMW buff was simply relieved to hear that Chris Bangle was not involved.

Under that svelte hood—which is really a giant hopper to hold shaved ice—lies the engine, a water-cooled 1.8-liter 4-cylinder Volkswagen industrial unit that runs on propane and puts out a modest 63 bhp. It's connected to a hydrostatic transmission that drives all four wheels through a pair of heavy-duty Dana axles, bolted directly to the frame.

That's right; there's no suspension. No shock absorbers or springs. Sort of like a TR-4 without all the pretension.

The business end of this rig is found at the very back, where ice is made.

Now, I had always imagined that a Zamboni ice resurfacer was some kind of mobile refrigeration unit, but it actually lays down a blanket of *hot* water on the ice. Here's what happens:

First, a sharp 77.0-in. blade scrapes the ice, then a horizontal screw forces the shavings to the center of the machine, where a vertical screw flings the shavings into the front hopper, or "snow tank." Wash water then floods the ice in a short, wide puddle that is immediately vacuumed into the tank and filtered for re-use. This cleans dirt and debris off the ice.

In the last stage, hot de-ionized water is fed through a wide cotton towel that lays down a new layer of water that freezes glass-smooth in minutes on the 18-degree-Fahrenheit rink. All cooling is done by refrigerant pipes buried in the concrete under the ice. The ice itself is only an inch thick, and the first layer is painted blue (using the machine) to give it that ice-blue Aqua Velva appearance.

Okay, so the thing reconditions ice. But this is a car enthusiast magazine. What's it like to drive? Well, let's clamber up the two diamond plate steps and see.

You sit high in this thing; it's kind of like sitting on the roof of a Cadillac Escalade, or behind the wheel of a 4$^{1}/_{2}$-liter Bentley. In front of you is a simple steering wheel with either a "suicide knob" or a "necker knob" (depending upon which high school you attended, and what kind of girls you dated) for easy turning, and there are two foot pedals: brake and hydrostatic control.

Debilzan tells us the rear brake drums are just there to meet some obscure industrial safety code. Real stopping is effected by the hydrostatic lever on the console. There's also a throttle lever with a pictograph of a rabbit (representing speed) at one end and a tortoise (representing hard-shelled terrestrial reptiles) at the other. (Ferrari, take note: With a legible rabbit emblem in his cockpit, Schumacher would be almost unstoppable.)

Next to the throttle are a bunch of levers too numerous to mention. Debilzan told me, "A Zamboni driver should never have his right hand on the wheel." You have to control the blade at all times, or you'll scrape all the ice off the concrete at both ends of the ice rink, where your clockwise spiral patterns overlap.

Complicated work, this ice making.

Riding with Debilzan on the rink during several acceleration runs, I soon discovered that the Zamboni leaves the line with alacrity (revved to its 3000-rpm redline), rockets up to 9.7 mph and then stays at exactly that speed for the full 200-ft. length of the rink. Those old myths that you will "black out" or "be unable to breathe" at speeds above 9.5 mph proved to be completely untrue. I was perfectly comfortable, once I got over the excitement, and felt no ill effects then or later.

Same with the skidpad. No head restraint collar needed here. The Zamboni turns well and holds its line, but at these sub-sonic speeds it generates a modest 0.17g. For being such a heavy vehicle, however, it stops quite quickly and you have to hang on to avoid being flung forward onto the ice and then shaved, augered and washed, which is not as much fun as it sounds, at my age.

There's an old saying that "Nothing hugs the ice like a Zamboni," and this proved quite true in our slalom runs. It blew through the cones at nearly full speed, its engine screaming at redline. Try that with your average SUV and you'll soon be on your roof, writing letters to Ralph Nader. This thing flat handles. Which it should, with no suspension.

The author and the Long Beach Ice Dogs take center ice, while mascot Spike (with dog collar foreground) strikes a pose.

Which brings us to comfort and amenities. Zamboni offers optional cupholders, coffee warmers, CD players and even a fully enclosed cab for outdoor use, but our test vehicle was a "stripper" model with none of these unnecessary frills. Still, it might be fun to have a CD player so you could crank up Warren Zevon's famous hockey song, "Hit Somebody," while you groom the ice.

So, is the Zamboni for you?

That's a question only you can answer, though it wouldn't hurt to get a second opinion, and then pay attention for a change.

Overall the Zamboni is a very focused vehicle, aimed at a narrow niche market. People who are ice rink owners or Canadians—or wish they were—will love it. But it won't appeal much to growing families and soccer moms, or even NASCAR dads, soul brothers or Sisters of Mercy.

There's not much luggage space, unless you dump the snow, and $50,000 is a lot of money for any sport ute with a top speed of only 9.7 mph. Also, you won't be allowed to operate the thing unless you are a trained Zamboni driver, so tight, curvy roads will require rapidly shouted instructions to your operator. A slight stammer here could prove fatal.

But if your thing is resurfacing tired old beat-up ice, there's hardly anything better on the market than the Zamboni 500. It's not only useful, but has entertainment value. As Charlie Brown once said, "There are three things in life that people like to stare at: a flowing stream, a crackling fire and a Zamboni clearing the ice." And as Paul Theroux observed in his novel, *The Mosquito Coast*, the ability to make ice is the defining trait of Civilization.

He didn't mention tripping, slashing, roughing, high-sticking or goalie interference, but maybe he's not a hockey fan.

Where Do Zambonis Come From?

Warm, sunny Paramount, California, that's where. In the late 1930s, Frank J. Zamboni, along with his brother and cousin, built a skating rink called Paramount Iceland and soon got tired of laboriously grooming the ice by hand. So Frank invented the first ice resurfacer, introduced in 1949.

In the early 1950s, skating star Sonja Henie bought one to take along on her world tour (talk about overweight luggage), and by 1960 the Zamboni made its Winter Olympics debut at Squaw Valley, California. Now they're everywhere, with more than 7500 machines operating at rinks and stadiums around the world. They're still made in Paramount—and still a family owned business, with a second factory in Brantford, Ontario, Canada.

May 2005

A Jaguar in Moose Country

An E-Type heads relentlessly north into Canada until the pavement ends

"So when are we going to take the E-Type on that long road trip we've always talked about?" I asked my wife Barbara a few weeks ago, suddenly looking up from my well-thumbed copy of Conrad's *Heart of Darkness*, a cheerful tale of death and mental derangement on a harrowing trip up the Congo River.

Barb looked at our kitchen calendar and flipped through the pages. "We have only six days left this summer with nothing scheduled," she said, "starting next Wednesday."

"That's five days from now...and we're leaving tomorrow for a three-day weekend with our friends...."

"Right."

"So we have one day to pack and get the car ready?"

"Right."

"I guess I'd better go check the tire pressures," I said, suddenly setting Conrad aside.

I concocted myself a tall pint of iced coffee in my favorite Guinness pub glass and went out to the workshop to contemplate our beige 1967 E-Type coupe.

We'd owned the car for five years, and I'd spent three dragged-out years doing a mechanical restoration from the firewall forward—new brakes, suspension bushings, clutch, transmission seals, cooling system, etc. My friend Denny Marklein's body shop had repainted the car in its original color; a place called Straight-6 in California had revamped the cylinder head; another friend, Steve Straavaldsen, had rebuilt the bottom end, and I'd installed and tuned it. At 65,322 miles, the beautiful but rather complex rear suspension was still stock and untouched. No funny noises, so I'd left it alone.

I'd done a recent oil change, but hadn't tuned the car, checked the valves, changed the points or set the timing in two years, but it was running fine nevertheless. No good reason not to embark on a big road trip.

So I set the tire pressures, checked the oil and water (fine), whacked the knock-offs with a lead hammer one last time, filled a small metal toolbox to the very top with every hand tool devised since the Bronze Age, slid it into the back of the car and shut the hatch.

"The car's ready," I told Barb.

"What are you taking for spare parts?" she asked.

"None," I said authoritatively. "Where would I start? If we take a spare water pump, the coil will fail; if I take an extra fan belt, we'll lose a wheel bearing. British cars always know instinctively what spares you're carrying. Better to take nothing."

Barb looked deeply into my eyes, as if trying to discern a rational brain buried somewhere behind them. "If you say so...." she said in a lost, slightly doomed, tone of voice.

"The car's ready," I said. "Now all we need is a place to go."

We got out our Rand McNally Road Atlas and looked at the big map of North America. It had been a dream of ours to drive the Jaguar out to Nova Scotia, but a quick glance at the map told us we'd need at least six days just to get there and back from Wisconsin—using the Interstate. Not a sports-car trip made in heaven.

We looked west at the Great Plains and the Rockies. Lots of flat, hot, empty space to burn up before we got to the mountains—which were, at that moment, on fire.

"Let's go straight north," I said in a moment of inspiration. "Drive until the paved road runs out. We've never been north of the

Boundary Waters Canoe Area in our lives. Let's take two-lane roads into Canada until we can't go any farther."

The more we thought about it, the better we liked the idea. It would be pure fun. No Interstates, big cities or boring roads. Just small towns, blue lakes and north woods.

"What's the last paved road straight north of here?" Barb asked.

We traced our fingers deep into Canada and found a lone secondary road that ran almost 200 miles due north from the Trans-Canada Highway into the green, empty center of Ontario. At the end of that road was a little town called Pickle Lake.

"Perfect," I said. "Let's drive to Pickle Lake."

"I wonder if they have a motel," Barb said doubtfully.

That night we drove to a Madison bookstore to shop for Canadian travel guides. Strangely, there was no mention of Pickle Lake in any of the five or six guidebooks we checked. Stranger still, there was almost no information on north Ontario in any of them. One book had the address of every gourmet coffee bar in Vancouver, but not one word about this Montana-size patch of rivers, lakes and trees.

"That's because there's absolutely nothing there," my friend Jim Wargula explained when I told him about this oversight. "Just mosquitoes and tundra."

We'd see. In any case, I liked the idea of driving into *terra incognita*. It made me feel kind of like Columbus, but with Lucas electrics.

Barb and I left on a Wednesday morning and pointed the long nose of the Jag north on small country roads. Ah, the aroma of old leather and wool, mixed with mown hay and barnyard-fresh air. It was a warm morning, so we opened the rear vent windows and turned on the ventilation fan.

The fan has two speeds, slow and fast. Slow turns at about 3000 rpm and "fast" kicks her all the way up to about 3002. The difference is almost discernible to the trained ear. Nevertheless, ventilation is surprisingly good in the Jaguar, and it takes a very hot day to make you wish for air conditioning.

We wended our way over to the Mississippi near LaCrosse, then headed north along the Scenic River Road. I made a brief stop at Bay City to visit the old fishing cabin I'd rented with two buddies one summer when we worked on a Burlington Railroad section crew.

Peering in the window of the unrented cabin, I noted that the same chrome-legged kitchen table and chairs were still there, from the summer of 1967. It was sitting around this table after work that I first heard "Light My Fire" by the Doors and "Somebody to Love" by the Jefferson Airplane. That fall, about two months later, I met Barb at college. Meanwhile, in England, our future Jaguar was coming down the assembly line, though we wouldn't buy the car for another 32 years.

We made our first fuel stop at Ellsworth, and it took 13.4 gallons and averaged 19.6 mpg. From there we drove into deep north woods and made it to Lake Superior by evening, cruising through the city of Superior and across a bay full of ore boats into the slightly superior city of Duluth, which has many swanky homes along the lakefront.

Just past Duluth, we turned onto the Scenic Shoreline Drive (shunning the Viewless Inland Truck Corridor) and checked into a lovely little yellow cabin in the pines at the Island View Resort. Dining nearby at the Bay Shore Supper Club, we sat at a window overlooking the lake and ordered a fish combo special that featured walleye, herring and lake trout. We also had a Manhattan, to celebrate a 489-mile day in which the tool kit stayed firmly closed.

I raised my glass in our traditional British car adventure toast: "So far, so good, as the sky diver said."

Just south of Grand Portage the next morning, an orange light flickered on our speedometer face.

"It's our low-fuel warning light!" I exclaimed to Barb. "It works! This is the first time I've ever run the tank this low. It's a miracle!"

The light came on at 280 miles, and we filled up with 2 gallons left in the 16¾-gal. tank. Just before the Canadian border, we stopped at Grand Portage, toured the replica fort and buildings on this famous gathering place for the French voyageurs, then headed for the land of the maple leaf.

We crossed the Canadian border station with a few polite questions about our intentions. "And what does the missus do for a living?" the guard asked me.

"She's a physical therapist who works with disabled school children," I said.

He smiled and waved us through. Apparently, this is a good answer in the age of drugs and terrorism.

I can't tell you why, but I always relax a bit when I drive into Canada. People are a little more polite, drivers a bit more mature, with less "attitude" (that dismal product of pop culture). On the surface, at least it feels like a calmer, more rational place. Also, you see far fewer cops everywhere. Maybe all those things are related.

The fort and forbidding falls at Grand Portage

We drove through rugged cliff country into the big town of Thunder Bay, then forked west on Highway 17, the famous Trans-Canada Highway. Pretty country, with big lakes, high ridges and vistas of green forest. The "towns," what there were of them, came along about every 60 miles—Raith, Upsala and English River. We drove into St. Ignace early in the evening with our low-fuel light fluttering and stopped at an Esso station. The tank soaked up 56.8 liters of fuel, for $56 Canadian. I didn't even try to figure out the mileage; it made my brain hurt.

St. Ignace was the town where Highway 599 turned straight north toward Pickle Lake for 292 kilometers, or "klickadoons," as Barb had taken to calling them, in deference to Canada's rich Scottish heritage. I asked about Pickle Lake at the gas station, and the very name seemed to make people laugh. "Lived here all my life, never drove up there," one man told me.

"Nothing there," the station owner said. "A mining town. Used to be. Fishermen go there. Why do you want to go up there with that beautiful car?"

"It's the end of the road."

"You can say that again."

A truck driver at the station had actually been there. "Couple of motels," he said. "Rough road, 10 km of gravel. Watch out for moose; they're everywhere. And bears. A friend of mine saw 10 black bears along the road last week. Don't leave any food in that Jaguar when you park it."

I was reminded of my favorite line from the old Jack Nicholson western, *The Missouri Breaks*, in which a Montana outlaw gang escapes into Canada. Riding through the dark forest, one of the outlaws looks suspiciously around himself and says, "There's things in Canada that eats your horse."

Or your Jaguar, apparently.

We found a motel just out of town on Lake Aginac and had excellent home-made lasagna at a restaurant attached to a gas station. As in Sweden, people this far north don't waste heated walls. A young woman at a nearby table looked out the window at our Jag and said, "What kind of car is that?

In our two days of driving, a lot of people had asked this question. To a car buff from the '60s, this is like seeing a Julie Christie movie and wondering who she is. But the very question is proof that generations change, time passes and, as Ecclesiastes says, "all is vanity."

The next day we hit the trail to Pickle Lake. It turned out to be a lovely road, sweeping back and forth among lakes and forests, rising and dipping over granitic ridges of the glaciated Canadian Shield, some of the oldest exposed rock on earth.

This landscape, with its conifers clinging to thin turf on scoured bedrock, reminds you that, only a little more than 10,000 years ago, the whole area was buried under a white sheet of ice. In geological time, this was just minutes ago. And here we were, rushing with our Jaguar for a quick look, mere seconds before the next ice age. Which makes you wonder about the long-term collector value of cars from the 1960s.

There were a few dips and bumps in the pavement, but mostly the road was in superb shape, with one short stretch of road construction. We passed signs for five or six fishing lodges, none of which had been mentioned in our guidebooks. Traffic was light to nonexistent, except for a few lumber trucks and—of all things—a red Miata.

At the little village of Savant Lake, we topped off our fuel at the grocery store/gas station. The car took 18.7 liters of fuel for $18.18 Canadian with 87 miles on our trip odometer, which worked out to be 39 furlongs per cubit, by rough estimate, squared. The woman at the checkout counter warned us to watch for black bears. "We've got some down at the dump as big as Volkswagens," she said.

I had a fleeting, ugly vision of VW Beetles breaking into a Jaguar for food.

About 85 miles later we passed through an Indian reservation and entered a town that appeared to have two names: Mishkeegogamang and New Onsaburgh. For the sake of multiculturalism, we decided to call it Mishkeegogamang, but pronounce it with a Peter Jennings accent.

North of this town, dead smack in the middle of nowhere, a cop suddenly stepped out of the forest and aimed a radar gun at us. I looked at my speedometer. I was going exactly 54 mph—the precise 90 km/h Canadian speed limit—because of dips in the road and a general relaxed sense of sightseeing.

The cop looked at his gun in disbelief, then got in his car and followed us. After about five miles, his lights came on and I pulled over.

A young and very polite officer from the Ontario Provincial Police leaned in our window and said, "Sorry to stop you, sir, but we don't get many exotic cars like this up here, and I just wanted to see if everything was okay."

"Everything's fine," I said.

"Good. May I see your registration, driver's license and proof of insurance, please?"

He looked over the papers and handed them back. "I just wanted to make sure you were the owners of the car," he said. I nodded and wondered if rings of exotic-car thieves were driving old Jags to Pickle Lake and then air-lifting them to Russia with de Havilland Beavers.

"This is a beautiful car," he said. "I don't think there's ever been an XKE on this road before. We have a T-Bird and an old Cadillac in Pickle Lake, but they never leave because of the rough roads." He then gave us advice on motels and gas stations and said, "Have fun."

As we drove away, Barb said, "I think he just stopped us to look at the car."

"If I saw this car up here, I would too. I'd set up a roadblock."

"I'll bet we really are the first E-Type on this road," Barb said, pleased with the idea.

"I think so," I said. "And, until disputed, we will stand by that claim."

A few minutes later, it started to rain so I turned on the Jag's small triple wipers, which work pretty well. Or at least they did until they suddenly stopped. As I strained to see through the rain-streaked windshield, we passed the city limits sign for Pickle Lake ("The Last Frontier") and the wipers suddenly started working again.

"Pickle Lake is the Lourdes of Lucas electrics!" I exclaimed.

When we finally got into town, Pickle Lake was better than described, but not exactly downtown Manhattan. It was a slightly tired-looking little town strung along the shore of a lovely lake, with quite a few people just standing around. There were a couple of motels, a decent airport on the hill above town and a fly-in fishing service with a Cessna 185 floatplane on the lake. A nice young couple who ran the charter flying service, Jason and Joy Bridle, came out to look at our car, and let us park it on their dock for photographs.

"This dock really looks like the end of the road," I said.

But it wasn't quite. The pavement actually ended about a mile north of Pickle Lake at the little village (i.e., convenience store) of Central Patricia. Here the pavement turned to gravel. A man at the store said the gravel road continued north for about another 100 km, and in the wintertime you could sometimes drive all the way to Hudson Bay on dirt roads and frozen lakes. That would be an adventure for another time, and another vehicle. A Morgan 3-wheeler, maybe.

We took some pictures of our car at the end of the pavement, and I noted that we'd gone exactly 1002 miles since leaving home.

As we retraced our route south in the rain, Barb got out her cellphone and called a fishing lodge whose sign we'd passed on the way up, Sac Bay Lodge Camp on Sturgeon Lake. The owner said he had a cottage available. "We're starving," Barb said. "Do you serve dinner?"

"Oh, I might be able to find something for you to eat," the owner said, in what Barb took to be a French-Canadian accent.

The camp turned out to be a beautifully restored log lodge and cabins on the lake, about a mile off the road, and the owner, Fred Wittwer, was originally from Switzerland. When we got there, he cooked us and one other couple the single best meal of the trip—smoked trout appetizer, a delicately sautéed walleye course, a salad with his own homemade dressing, steaks cooked on the grill and his own special after-dinner drinks made with coffee and some bliss-inducing mixture of spirits. Without really trying, we'd stumbled on one of the prettiest spots in Canada and the best chef north of Miami.

The next morning we slithered south on the damp curving roads toward the Trans-Canadian. The E-Type was perfectly suited to these roads and great fun to drive. With all our luggage and extra tools in the back, it felt a bit soft at the rear over big dips in the road, but was otherwise both supple and composed, with precise steering and nice feedback in the endless curves. The 3.54 diff, which sometimes made the big, torquey six seem unnecessarily busy on Interstates, put the car right in its element on winding secondary roads, with instant acceleration and great throttle response. It was whippy and quick.

Which was nice on Highway 502, the road south from Dryden, possibly the prettiest stretch of road we saw in Canada. With big, sweeping turns around marshes and lakes, it threaded its way from one isthmus to another, through much exposed granite and tall pines. Good GT car country, like the road to Pickle Lake, but with perfect pavement.

And on this road, we finally saw a bear—a large black bear standing idly by the side of the road. As we stopped to take pictures, a trundling motorhome came around the corner, holding up a parade of about 60 cars. Sight of the bear caused near pandemonium in this group—locked brakes, honking, etc. The bear fled and so did we.

We crossed Rainy River and the border at International Falls, where warning signs on the highway instantly switched from moose to deer. I don't know why, but I get the impression that no moose would ever consider leaving Canada and crossing the U.S. border. It's just a matter of style and tradition. In any case, we'd seen no moose in either country.

Before heading back into Wisconsin, we had one last travel goal, and that was to visit Hibbing, Minnesota, Bob Dylan's home town.

Biographies of Dylan have made Hibbing sound like a desolate, half-deserted mining town where the Greyhound sign creaks in the wind and young poets go mad from ennui in the wintry darkness. In reality, it's quite a lovely city with a nice downtown, shaded streets with grand old homes, great-looking public schools and proud civic architecture of brick and stone. It appeared that much iron-mining wealth had been spent on infrastructure.

With directions from a gas station, we found the old Zimmerman (Dylan) home on 7th Avenue in a nice middle-class neighborhood near the high school. It was a blue two-story, with interesting, Indian-style trim around it. As I was taking a photo of the Jag in front of the

At left, a visit to the boyhood home of Bob Dylan in Hibbing, Minnesota. Above, current owner Greg French and his pal "Maxie"

house, a Frito-Lay truck pulled up, and the driver jumped out and introduced himself.

He was Greg French, a friendly, articulate man and owner of the Bob Dylan boyhood home, just coming home from work. We apologized for standing in front of his house, but he said he didn't mind at all, and had us sign a visitor book. "We get people from all over the world," he said. French told us he'd lived down the street, needed a larger house for his growing family, 14 years ago, and found out the Zimmerman house was on the market. "We paid more than market value," he said, "but I thought the house was worth preserving."

We thought so too. Along with the iron ore that won World War II and put America on wheels, Bob Dylan was certainly Hibbing's most famous export. Barb and I had tickets to see him in Madison, two weeks after our trip.

We left Hibbing late in the afternoon and drove 200 miles south to Stillwater, on the St. Croix River, to stay overnight with our old friends, Bruce and Linda Livermore. Linda was gone for the weekend, but Bruce ran out and got us a big pizza. The Livermores' kids, Megan and Ross, were gathered around the piano in the living room singing Beatles' songs (superbly, I might add) with a bunch of their equally talented high school friends.

The old Jaguar, standing nearby in the garage, probably felt right at home. Especially since it was sharing garage space with my old Lotus Seven and Bugeye Sprite, which Bruce now owns. It was a nice little reunion of stray British spirits—"in some corner of a Minnesota garage that is for ever England," to irreverently paraphrase Rupert Brooke.

The last day of the trip we meandered lingeringly on back roads through the green hills and red barns of Wisconsin's unglaciated Driftless Zone, through my erstwhile home town of Elroy and back to our present home, south of Madison. One of the most beautiful parts of the drive, in our own back yard.

We arrived in the driveway having gone 2029 miles. The Jaguar averaged 19.1 mpg and used two quarts of oil on the trip. Except for the self-healing wiper problem, the car had run for five long days without missing a beat. I'd never opened the toolbox. It had been, truly, a grand tour.

Which is what the E-Type was built for. There's something to be said for driving an old English sports car that was actually designed with continental touring in mind, intended to go from London to Zurich or Sicily, rather than just running over the hill for a quick pint at the village pub. It's a car with seven-league boots and deep reserves of performance.

And it still works well on modern roads. Many current economy cars will cruise just as easily and comfortably at 70 mph on the highway, but very few of them will lunge effortlessly to 100 mph (or much, much more) right now, when you need to pass that sluggish motorhome or gravel truck in a hurry, then drop back to a smooth, muted growl, full of latent possibility.

It's this two-stage personality that makes the E-Type coupe such a good companion on a long trip. It's fast and fine-handling, but also comfortable, with compliant suspension and lots of luggage space; civilized, yet burning bright, like Blake's famous jungle cat. The old Jaguar slogan of "Grace, Pace and Space" had it exactly right.

As we unloaded the car that evening, I found myself quietly humming a song written long ago by a Canadian named Neil Young, about an old car he owned in north Ontario. It was called "Long May You Run." It has a beautiful line about "trunks of memories, still to come."

May 2005

A Moveable Beast

Our friend Jay Leno builds himself a 1966 Olds Toronado sleeper with just a few extra horsepower

"I like to think of this as an ideal Grand Touring car for guys who are older than 50," Jay Leno said as we blasted up a winding road into the hills above Burbank, California.

I put my foot down for a moment and the big Toronado stepped out slightly at the rear, hooked up with all 700 bhp and threw us toward the next corner with casual brutality. The sound was magnificent—a combination of mellow, deep roar and high-pitched rip, like an angry bear opening your pup tent without using the door.

As we braked hard and pitched into the next corner with distinctly un-1960s' flatness and precision, I tried to think why Leno thought this car should appeal more to guys older than 50—which I am—rather than, say, the average 25-year-old.

Grim fatalism? A palate jaded by too much speed and hard living? Compensation for lost youth? Gum recession?

No, there was more to it than that. A couple of things came to mind.

First, we older types grew up with cars that looked like this Toronado—clean, neatly sculpted shapes from the Bill Mitchell era at GM—and, second, our expectations of how a car should handle, stop and ride have been raised slightly in the past 40 years.

Leno had put the old look and new performance together, essen-

Right, Leno with his framed cover of R&T's November 1965 issue, featuring the 1966 Toronado, his personal favorite. Top, Jay gives Egan some pointers before hopping behind the wheel of this wickedly fast rear-drive Olds.

tially folding a 1966 Toronado, a killer engine and a C5 Corvette chassis into a single car. An appealing combination for those of us who can remember when Steve McQueen was Josh Randall, but can't find our reading glasses.

From that vantage point, Leno's car was an interesting project, certainly. But people often slide big, modern fire-breathing Chevy drivetrains under old car bodies. Too often, some would say. Hot rods sometimes seem more common than unmolested cars with their original charm intact. So why had we driven all the way up to Burbank on a nice winter morning to drive this particular example?

Curiosity, mostly, along with the usual subliminal desire to see what else Jay has in his garage, which is always fun. How could we resist?

Most of America knows Jay Leno as the host of *The Tonight Show*, of course, but those of us with our priorities straight think of him mainly as a consummate car and motorcycle buff with a good job, a guy who spends most of his spare time on interesting mechanical projects. Such as the Toronado.

The gradual construction of this car had been much covered in the automotive press during the past year, but no one had gotten behind the wheel until Leno called the R&T offices a few weeks ago and

asked if we'd like to drive it. He described the Toronado as a really nice GT car, a useful daily driver and "a pretty good match for the new Bentley Continental GT I drove in England last month."

Hmm. That was quite a claim. The Bentley Continental GT is a real piece of work. It's an elegant and ultra-civilized, high-tech all-wheel-drive car with a twin-turbo 6.0-liter W-12 engine that cranks out 552 bhp and a massive 479 lb.-ft. of torque. It costs $156,000. If Leno—who knows quite a bit about cars—thought he had some kind of vintage American *Doppelganger* for the Continental GT, maybe the Toronado was worth a look.

So on a sunny Saturday morning, photographer John Lamm and I drove from Newport Beach up to Leno's famous Big Dog Garage. This restoration shop/garage complex, with its collection of some 80 cars and an equal number of motorcycles, is housed in five neat industrial buildings at the Burbank Airport. When we got there, Leno was adding some very dark and viscous oil to the engine of a Doble steam car.

"Lifting the hood is like opening your refrigerator and finding a nuclear reactor instead of milk and orange juice."

He greeted us cheerfully, wiped his hands off with a rag and led us over to the gleaming Trumpet Gold Toronado, which was poised in the open garage door. It looked quite stock, except for wheels and tires that seemed to fill the wheel wells.

"I duplicated the look of the stock wheels, but eliminated the cheesy-looking original hubcaps," Leno explained. "Each wheel was machined from a 400-lb. block of aluminum, and the tires are 17-in. Bridgestones with red stripes added to make them look like the old

Uniroyal Tiger Paws."

He opened the doors and we looked at the interior, which has a stock dash and steering wheel, with black upholstery that Leno describes as looking correct for the period, but not stock. The only interior giveaway to anything funny going on under the sheet metal is a small row of intercooler toggle switches on the dash. This is not normal.

Neither are the exhaust-gas temperature and boost gauges mounted in the stock air-conditioner outlets. A well-hidden Vintage Air a/c system supplies cool air from under the dash.

If the car looks pretty stock from the outside, it certainly doesn't when the hood is raised. No greasy old cast iron here; there's so much polished aluminum, the engine compartment gives off its own light. Lifting the hood is like opening your refrigerator and finding a nuclear reactor instead of milk and orange juice.

"So a 3.43:1 rear end was installed, reducing top speed to a more practical 247 mph."

The Toronado's engine is a 425-cu.-in. V-8—same displacement as the original Olds unit—flanked by turbos and a pair of intercoolers built into the fenderwells. Leno said most of the parts came from GM Performance Parts, but the engine was assembled by Lingenfelter Performance Engineering. Installation and all chassis work were done by Leno's shop foreman, Bernard Juchli, and race-car builder Jim Hall (no relation).

As delivered, the engine produced 1172 bhp and 1000 lb.-ft. of torque, but that was deemed too much for the moderately sized tires to handle, so the wastegates were dialed back to

a mere 700 bhp and 925 lb.-ft. of torque.

It would have been fun to see Leno and his crew try to put all that power to the ground through an original Toronado-style fwd system (torque steer, anyone?), but that might be a project better suited to NASA.

Instead, the front and rear suspension (including brakes) of a C5 Corvette were grafted in, with a subframe of 4-in. steel channel holding everything together. The rear-mounted Corvette 4-speed automatic transaxle uses heavy-duty pieces from a Cadillac Escalade transmission, and the car has full ABS and traction control.

With the stock 3.15:1 Corvette rear end, a theoretical top speed of 270 mph was charted, but this was deemed impractical for daily commuting to the Burbank Studios, so a 3.43:1 rear end was installed, reducing top speed to a more practical 247 mph.

"This level of acceleration is so much fun, it's hard not to laugh out loud every time you get on the gas pedal."

There's a 22-gal. fuel cell in the back (with plenty of trunk space remaining for luggage and CD player), with a pair of fuel pumps. At the front, a large transmission cooler sticks down below the front apron.

If this all sounds like a project with a high Frankenstein factor, it certainly doesn't look that way from underneath. Everything is clean, neat and beautifully routed, like something that just came off the factory assembly line, only better.

But, of course, Jay's Toronado doesn't share much DNA with the original factory item.

That car was a GM technical tour de force in which Oldsmobile engineers somehow managed to make a big, powerful, longitudinally

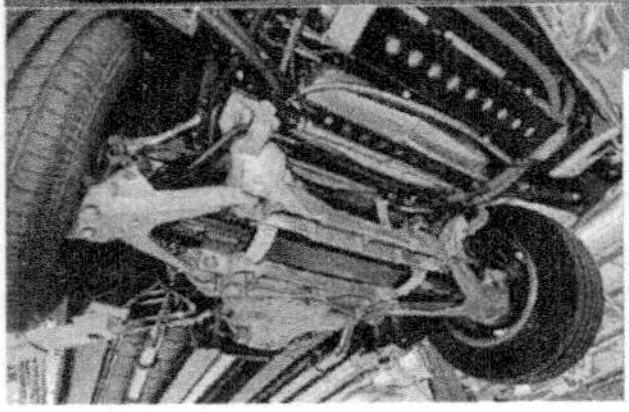

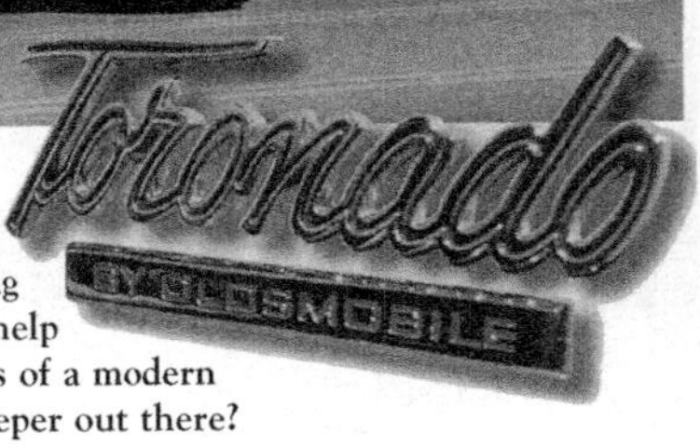

Jay's car looks pretty stock, but C5 Corvette aluminum suspension, front and rear, along with Bilstein shock absorbers, help give the Toronado the dynamics of a modern sports car. Is there a better sleeper out there?

mounted V-8 feed power to the front wheels without twisting off the axles or steering the thing sideways into the nearest ditch. Besides being technically sophisticated, the Toronado was a really good-looking car. Or at least I always thought so.

And so did Jay Leno.

"I always loved the looks of this car," he says, "and I wanted a '66, which I thought was the best year." He asked some friends who supply old cars for the movies to keep their eyes open, and they eventually found a derelict old Toronado for $800. A few thousand highly skilled man-hours and Lord-only-knows how many thousands of dollars later, and it was transformed into the car before us.

Time to drive.

Leno fired up the car—which has a wonderful, throaty rumble at idle and emits the sweet smell of racing gasoline—and we headed toward the nearby foothills. After a short demonstration of the basic mind-boggling nature of the car, he pulled over and let me drive.

The Toronado stops, starts and parks just like any other 1960s' sedan with an automatic transmission. Fire up, select Drive and move out. It's not abrupt, cranky or difficult to operate in any way. Put your foot down, however, and the illusion of docility fades quickly. The car leaps down whole city blocks in one throat-clearing cough of power, and on the highway it just flat moves out, flinging huge segments of pavement into your rearview mirror. It's a simple case of all the power you could ever want, limited only by tire traction. This level of acceleration is so much fun, it's hard not to laugh out loud every time you get on the gas pedal. Driving it around is like some victimless form of juvenile delinquency.

Steering is easy and neutral, ride is civilized and the brakes are strong and linear. What the heck; it's a Corvette chassis. Kind of.

But just beneath the rich, deep rumble of the twin glass-packs you are always vaguely aware of the hustle-bustle of relentless drivetrain management—fuel humming through its return lines at 90 psi, waste-gates fluttering open and closed, the induction system inhaling air by the cubic yard. The whole drivetrain gives the impression of being a live thing, restrained and determined to get out, like a pit bull locked in a closet. It's contained for now, but not happy about it.

We turned south off our canyon road and headed out onto the freeway. Here too the nonchalant but overwhelming power delivery makes the car a lot of fun to drive. See an open slot in the flow of traffic and you're there, no waiting. Need to get off at the next exit? Getting ahead of that garbage truck before you get there is no problem. It's wish-fulfillment, fighter-plane power. Relative to other traffic, you feel like the only live person in a wax museum.

"What does this thing get for fuel mileage?" I shouted to Jay, after briefly gunning the car up to 110 mph on an empty stretch of road before I ran out of room and slowed back down, reversing the g-load on my neck muscles and jowls.

"About 13 mpg," he said.

I shrugged. Not bad. About the same as my old 1963 Cadillac.

We headed back to the Big Dog Garage at an easy, mellow cruise, interspersed with random blasts of lazy yet manic speed. In a strange way, the Toronado reminded me of a Marshall JCM 900 guitar amplifier I used to have. The gain control knob went to 20, instead of the usual 10. I never cranked it up that far—5 was plenty loud and overdriven—but it was nice to know you could. At low volume, the amp had a pleasant tone, but a quick flick of that knob could make your eardrums bleed.

That's Leno's Toronado. It goes to 20, but you can drive it at 5 and enjoy the music.

So does all this make the Toronado a real GT car? A reasonable stand-in for the likes of a Bentley Continental GT?

Sure, why not? Hard to tell about long-term service and reliability, of course, but there's really nothing else to dissuade you from traveling across the country with this immensely powerful beast. Jim Hall tells me that, just after our visit, they installed an alcohol and water injection system, so the car will now run on unleaded pump gas, rather than racing fuel.

So the Toronado might make a pretty good daily driver, too, if you could learn to relax and quit humiliating those less fortunate than yourself, such as the owners of large German tuner cars with vanity plates and privacy glass. Or Bentley owners, for that matter.

Which is as good a reason as any for having a 700-bhp Toronado. It's a serious car with a good sense of humor.

July 2005

The Road to Sebring

Cruising the Gulf Coast, at long last, in Panoz' hotest version of the Esperante

Back in that innocent era just before the Beatles discovered acid and were still cheerful, there were essentially three stimulating things a college kid could do over spring break: 1) Go to Daytona Beach and try to meet girls who looked like Paula Prentiss and Sandra Dee; 2) Drive to Florida for the 12 Hours of Sebring in a cool sports car and camp out with your buddies; or 3) Go home and work in your dad's printing shop for $1.50 an hour so you could pay for your books and tuition.

Following my dad's heartfelt advice, I always did the last. After all, he was wiser than I, and outweighed me.

But one of my fellow journalism students at the University of Wisconsin, future R&T photographer and writer John Lamm, fared a little better than I did. He actually managed to finagle a *journalism grant* that paid him $300 to go to Sebring in 1965 and write a story about it. With no deadline!

This, my friends, is audacity.

He drove to Florida with a buddy in a Sunbeam Alpine and they "camped" in the car during one of the rainiest 12-Hour races in history. They sat bolt upright all night in those bucket seats, trying to sleep with rain blowing in around the edges of the convertible top.

Proving, I guess, that my dad was right about the benefits of stay-

ing home and printing.

Still, I've always wanted to go to Sebring for the 12-Hour, and I've long been envious of John's youthful adventure, though certainly not to the point of smoldering resentment, as my therapist has so often suggested. It's just one of those things a car buff should do in this lifetime, like seeing Monaco, Le Mans or Indy.

So when the folks at the Panoz car company offered me a chance to drive a new Esperante GTLM roadster from one of their distribution points in Houston, Texas, all the way around the Gulf Coast and to be their guest at the 12 Hours of Sebring, I felt the immediate relaxation of some great knotted-up tension from an unrequited dream. Off to Sebring at last, in a real 2-seat sports car—and with a buddy, no less.

The buddy, of course, was John Lamm. As soon as he heard about this trip, he volunteered to come along and take pictures. "I'd love to go back there again," he said, "as long as I don't have to sleep in the car."

I assured John of a condo near the track—and 340 more horsepower than the Sunbeam Alpine, as well as a better quality convertible top—so he flew into Houston from California and I came in from Wisconsin. We were met at the airport by Dan Panoz and his staff.

Panoz turned out to be a personable guy with a great sense of humor, abetted by his charming red-haired Irish wife, Melanie, whom he met when the Panoz family lived in Ireland. His American entrepreneur father, Don Panoz, invented the transdermal patch, among other things, and set up a pharmaceutical company in Ireland.

The family now owns the Sebring, Road Atlanta and Mosport racetracks, as well as a couple of hotels, vineyards and a car company founded and run by his son, Dan, that builds sports cars and Indy-winning Panoz IRL race cars at a factory 55 miles northeast of Atlanta. They are a lively bunch, in the best can-do American tradition.

After lunch at a nearby restaurant, Dan showed us around our new black sports car.

Panoz produces three versions of the road-going Esperante sports car. The standard model and GT both have naturally aspirated 305-bhp Ford SVT V-8 engines, but our little chariot was the supercharged GTLM version (the LM is for the American Le Mans series, created by the Panoz family to promote American endurance racing). This car features the longer and lower aerodynamic nose and bodywork developed for racing, as well as a 420-bhp version of the 4.6-liter Ford SVT 4-cam engine.

The body is all aluminum, and the subframe and floor structure are a combination of carbon fiber over extruded aluminum, with alloy steel front and rear subframes for the fully independent front and rear suspension arms. The car weighs 3384 lb., uses a 6-speed Tremec T56 transmission and has an MSRP of $121,325. Panoz lists its 0–60-mph time as 4.2 seconds, with a top speed of 180 mph and EPA mileage of 16 city/22 highway.

I climbed behind the wheel for a brief familiarization drive with Dan Panoz. Comfortable leather seats with good inflatable lumbar cushions, nice light clutch, precise, short-throw shifter, effortless drive ability at low speed with ever-more-thrilling acceleration as revs rise and the supercharger begins its muted scream. The view over the hood is somewhat Boxster-like, and the car feels about the same size, but has considerably more kick at the rear tires. Wind flow with the top down is serene and pleasant. Dead flat cornering, responsive steering and civilized ride, with just a little orbital motion of the chassis over undulating road surfaces.

The top goes up and down with a button and two manual corner latches. We opened the trunk and found plenty of room for John's big camera pack and our two duffel bags. None of the usual irritants, blind spots and compromises that so often go with hand-built performance cars. Except for a Blaupunkt radio/CD player so complex even Dan Panoz said he couldn't figure it out. We never did master it on the trip, so it was a music-less drive.

Waving good-bye to our Panoz friends, we hit the road late in the afternoon, escaping rush hour toward Galveston and the Gulf. After a long cold winter in the north, the warm sun beaming down into a convertible was almost medicinal.

Having never driven around the Gulf Coast before, everything I knew about Galveston was contained in a Glen Campbell song and in the vague warnings of friends who said it was "a mixed bag" or "somewhat seedy."

We crossed over the bridge onto Galveston Island, drove down the coastal sea wall looking for a hotel, and, contrary to those warnings, I found the town quite pleasant. It reminded me of the Old Florida of my youth, full of small, family-owned motels with names like Sea-Spray or Ocean Breeze, places where you can park right in front of your room and listen to the palm trees rustle around the pool while waves lap at the sandy beach. I much prefer resort areas like this to the mega-hotel and golf complexes that seem to dominate modern beach construction.

We found a nice old motel right on Seawall Road, and walked down to Gaido's Seafood Restaurant and had a superb meal of oysters and crab cakes from Galveston Bay.

Back at the motel, a 3 a.m. party suddenly broke out in the room next door, with about a dozen young spring-breakers whooping it up and drinking shooters (or something) at the count of 3, over and over again.

"This can't last long," I mumbled to myself, "unless they're drinking milk."

It took about 45 minutes for the last drinker to fall down with a muted thud and go silent.

In the morning it was cool and rainy, and John and I were psychologically unable to leave town without first visiting the nearby Lone Star Flight Museum. We're both World War II airplane buffs, and I had to pay a respectful visit to the museum's P-38 Lightning, whose fuselage waist scoops and twin-boom rudders were borrowed by Harley Earl for the styling of my 1953 Cadillac. In the museum, we met a former military pilot named Bob Searcy, who flew Skyraiders in Vietnam in support of ground troops. As John and I were both ground troops in that conflict, we all had a mini veterans' reunion under the nose of the museum's Skyraider.

I couldn't help noticing that we were all beginning to look like WWII vets used to.

"We crossed swampy channels and salt marshes into Louisiana, and the landscape immediately changed from low scrub into classic Old South."

John and I drove onto a car ferry called the *Roy Stoker* and crossed Galveston Bay toward Port Arthur (most famous in my mind as the home of Janis Joplin), past oil rigs, ship-repair dry docks, fishing fleets and agri-business equipment yards, all half rusty. This part of the Gulf feels like the industrial underbelly of America, but it has an honest, hard-working feel to it that I like. It's a place where material objects are still being moved around, rather than just "information."

We crossed swampy channels and salt marshes into Louisiana, and the landscape immediately changed from low scrub into classic Old South, with huge live oaks hung with Spanish moss, shaded antebellum homes, lazy bayous and French Cajun names on the storefronts. In Vermillion Parish, Highway 82 swung north toward New Iberia, where John and I wanted to stay because we're both big James Lee Burke fans and all his Dave Robicheaux detective novels are set in this town.

Also, my good friend Lyman Lyons grew up here and I've heard so much about the place, I feel as if I lived there in another life. We cruised down New Iberia's beautiful old Main Street past The Shadows plantation home and, sure enough, there was the Evangeline Theater, where young Lyman once saw Roy Rogers matinees and the original *King Kong*. Behind Main Street was the dreamy, shaded Bayou Teche where Lyman used to fish and watch the river barges motor by. His descriptions of the place had been far too modest; this is a beautiful town.

John and I dined at a place called Clementine's on Main Street (superb crawfish pie), then found rooms at some soulless modern box out by the Interstate. It was too dark and rainy by then to search for more charm.

We left town on Highway 182 in the morning, passing signs for Avery Island, home of Tabasco sauce—most of which is spattered all over my Paul Prudhomme's *Louisiana Kitchen* cookbook—and hit the Interstate south of Jeanerette. At Lockport, we took a long side-trip down Highway 1 to the Gulf. This road follows Bayou Lafourche, said to be an ancient path of the Mississippi, past endless shrimp boats tied up in front of their owners' houses and out to Grand Isle, where hundreds of vacation homes stand on stilts, awaiting the next hurricane, their sacrificial (or getaway) cars parked beneath.

Backtracking to the Interstate, we cruised into New Orleans late in the afternoon, heading straight into the *Vieux Carre*. We found a couple of rooms at the charming old Cornstalk Fence Hotel on Royal Street—a place I'd discovered on my first motorcycle trip to the Big Easy in 1978—and then walked to K-Paul's Kitchen for another infusion of the kind of food (seafood *etouffee*) that always makes me wonder why I don't live down here.

At the hotel, I told the desk clerk that my sink didn't drain and he said, "Yes, the drains are old and can't keep up with the water pressure from our new pipes." When I mentioned that my room was a little hot, it was suggested I put my phone book on top of the heat register.

Everything I like about New Orleans may be found in those two comments. It's the polar opposite of, say, Prussia. Magnetic south.

The next day we had to Make Tracks, or miss Sebring, so we had a couple of quick French-roast coffees and *beignets* at the famous Café du Monde for breakfast. This place used to be staffed by haughty but courteous old men in black suits and white aprons, but no more. Rudeness and chaos prevail. Quite a change in tone, and another era gone.

We left town and blasted on the Interstate for a while before dropping back down to the Gulf at Pensacola. Highway 98 was crowded with traffic, hotels and new condos from there to Panama City—a zone of endlessly repeated franchises that John calls "the architectural sound loop"—but from Apalachicola east the Gulf was wild and beautiful, with small towns, fishing resorts and small seafood joints. My kind of place.

With darkness falling, we hummed into the deep, almost unpopulated forests of the Florida Panhandle, passing scores of deer. Unlike Wisconsin whitetails, which can be counted on to leap suicidally in front of your car, these Florida deer just stand along the roadside and watch you go by. They're either smarter, or too lazy to move. Or both.

The Panoz was running great, though the weather had been too cool and drizzly to put the top down since Galveston. Top up, it was quiet and snug, an ideal all-purpose road car with a civilized ride on dull roads, inspired handling in the occasional curves and enough power to dispose of rolling roadblocks—instantaneously—wherever you found them. With traction control off, the rear end could be induced to step out nicely.

The Panoz also attracted considerable attention at gas stations, and a surprising number of people (mostly young men) knew exactly what it was. Many thumbs-up and "Nice car!" comments. Also, the price surprised no one. People seemed almost disappointed it wasn't more. When you can't afford a particular car, I suppose the next best thing is shock value.

After our long drive through the enchanted forest of deer, owls and big trees that looked as if they could talk, we pulled into Perry, Florida, late at night and got the last two motel rooms in town, that, it must be said, were cleaner than you'd expect.

We rose early the next morning and headed south for Sebring on

Highway 27. Just 20 miles north of the racetrack, after four days of dark gloomy weather on the road, the clouds suddenly parted, the sun came out and the day turned gloriously warm. Birds sang and the smell of orange blossoms wafted through our car.

We turned into the track and found Dan and Melanie Panoz at their Chateau Elan Hotel and Spa at Corner 7. We walked out on the back deck to watch a couple of the Friday afternoon support races. Sebring at last.

Dan drove us over to the Panoz paddock, where team manager Ed Triolo gave us a tour of the two Panoz GT-2 cars, which had qualified 1st and 5th in class. They were heavily modified versions of our own car, with about 80 more bhp. Mercifully, the Blaupunkt radios had been deleted.

We also met the patriarch of the family, Don Panoz, a man who somehow mixes sparkling good humor with quiet sagacity, and also chain-smokes cigarettes—no doubt with some of the money he made off the nicotine patch. When another magazine journalist pointed this apparent contradiction, he said, "Do you think the inventor of the condom always used one?"

We stayed in a condo near the track that night, rather than bolt upright in our car, while parties, mayhem and drinking raged nearby in the notorious Green Park "Zoo" of the infield. John and I briefly considered touring this moral Inferno, but decided we'd already

The Esperante trip took Egan and Lamm past Galveston ferries, top, barges in New Iberia, middle, and to the Lone Star Flight Museum left.

lived through such things many times and had nothing more to learn.

In the morning, the 12-Hour race started at 10:45, with 38 cars howling toward us down in Turn 1, two LMP1 class Audi R8s swooping around in the overall lead. The Panoz cars were successfully defending their positions, holding off swarms of Porsche 911s. John went off to photograph various corners, and I set off for a hike of the full track.

I was surprised how many good viewing spots there are on Sebring's corners, with fences close enough to see the drivers at work. The track facilities are in remarkably nice shape, and Dan Panoz told me the challenge with Sebring ownership has been not to make it too nice—to keep some of that famous car-testing roughness in the track, some of the rusty old airplane hangars on the perimeter and the general party atmosphere. Sebring is supposed to be just a little bit seedy and wild.

Most of the infield crowd, however, seemed to be families and race fans in motorhomes and vans, lining the fences and paying close attention to the race. In Green Park, of course, you still had the traditional, wacky encampments and scaffolds of the humorously deranged—The Dodge City Dump, Mudsharks, Bahama Fun & Fish Club, etc. But even these guys were watching the race, albeit through a kind of stunned hangover fog. The infield also had a Corvette Corral and Porsche Park, for those few hearty souls who still drive their sports cars to the races.

By late afternoon the Panoz cars had been in and out of the pits with a few problems. One had a blown tire that damaged the bodywork, and the cars were shearing drive pins at the rear wheel hubs—a problem they'd never had before, of the "bad batch" variety. The No. 51 car, nevertheless, got patched up and back out there as the big crimson sun sank behind the orange groves.

I suddenly realized it was the growing darkness that made this place feel like Sebring, rather than some ordinary afternoon sprint race. Just as the shadows get long, when you'd usually be folding the lawn chairs and heading for home, you've got four hours of racing left, and things are really heating up.

At dusk, brake rotors on the big Astons and Corvettes begin to glow a brighter orange and their exhaust pipes flicker tongues of flame against the dark track. Everything turns black and orange, like Halloween. There's something about racing combined with darkness that makes it all seem wilder and more primal, as if there might be werewolves out there on the track now, pupils dilated, watching the circuit in pencils of yellow light. It makes the cars look angry and dangerous, and the drivers more lonely in the glow of their instrument lights, not as connected to the crowd.

At every corner, headlights slash though the track smoke like lasers, and there's a lot of smoke in the air—brat smoke, steak smoke, campfire and charcoal smoke, tire smoke and dust from too many clipped apexes. The air actually smells good, like food mixed with speed and adventure, and there's still that faint undertone of orange blossom.

The crowd—the real race crowd along the fences—hasn't diminished. If anything, it's more sober and intent, packed in tight and three-deep around all the corners. The final hours of darkness are adrenaline time, for fans and race teams.

When it's finally over, the Audis have come home, 1 and 2 overall in the top Prototype class. One of the beautiful Aston Martins has won GT1, ahead of a gallant Corvette effort for 2nd and 3rd, and a Porsche 911 has won GT2. The remaining Panoz, after much garage time lost to hub repair, has still managed to finish 7th in GT2.

As celebrations gather steam in the pits, John Lamm and I suddenly realize we are desperately tired, after 12 hours of walk-

An American roadster in the French Quarter of New Orleans.

The factory Esperantes, which had mechanical troubles at Sebring.

ing and watching. I can only imagine how the drivers and crews feel. Twelve hours is a long time to race, a long time to keep a car running at this crazy pace. It almost defies reason that any combination of technology and metallurgy could survive such a prolonged beating. You have to stand by the track all day and into the night, listen to the ear-splitting wail of engines pushing their headlights down the back straight at 176 mph and then watch the cars jostle and hustle each other into the hairpin, lap after lap, with a thousand whooping downshifts, to understand.

As I finally do, after all these years.

And, I must say, driving a sports car to Sebring and watching the 12-Hour over spring break beats job printing, hands down, despite the obvious appeal of money for books and tuition.

November 2005

The Star-Crossed Sports Car

Celebrating the 50th anniversary of BMW's sublime 507, with a drive from Verona to the Villa d'Este

The whole thing was quite Shakespearean. We were two gentlemen, not necessarily of Verona, but certainly wandering around in Verona, looking for the famous balcony from which the fictitious Julia spoke to the imaginary Romeo. Fellow motor journalist Ken Gross and I found this chosen courtyard on a side street, packed with tourists and festooned with hundreds of lovers' notes and favorite quotations tacked to the old timbers beneath the balcony.

Ah, the power of art. Shakespeare could have picked any city in Italy as a setting for his play, but he picked Verona and now you had tourists from Tokyo, Cleveland and Murmansk getting off buses with stars in their eyes, 410 years after the play was written.

Unmentioned by Shakespeare was a good pizza joint just around the corner, but Ken and I had noted the place and retired to it, slipping away from the crowd. "I'm beginning to identify more with Falstaff than Romeo these days," I said, ordering a large anchovy pizza and a beer. "Maybe it's an age thing."

Timeless drama was all very well, but we were in Verona with a handful of other journalists to celebrate the staying power of another kind of art entirely.

BMW had invited us there to celebrate the 50th anniversary of its 507 sports car—as well as its more luxurious and civilized companion, the 503—and to drive a gaggle of these cars from Verona to the shores of Lake Como for the *Concorso d'Eleganza* at Villa d'Este. Someone, as they say, had to do it, or this essential work would never have gotten done.

So the next morning we met at our departure point, a resort just outside of Verona. BMW had lined up five cars from its museum—three 507 roadsters in red, black and white, and a pair of 503s, a blue cabrio and a maroon coupe.

I walked around the red 507 roadster—complete with Rudge wheels—that Ken and I would be co-driving on the first leg of the two-day trip and realized that I'd not only never driven a 507, but had never actually seen one driven on the road.

BMW made only 253 of these cars between 1955 and 1959, so there weren't a lot of them floating around. Elvis had owned one, and so had John Surtees and the Aga Khan, but you didn't see those guys very often in the daily flow of traffic. Sales of the 507 were disappointingly slow, and it was only a brisk business in the tiny, plebian BMW Isetta that kept the company solvent during the mid-1950s.

Why were so few sold?

Well, first, BMW management had made a decision in the late '40s to go for the upper end of the market, and the 507 was breathtakingly expensive for 1955, with a price of $8000 [$8988 list price, Road Test, October 1957 *Road & Track*], or about two Cadillacs' worth. And this was a time when Europe—and particularly Germany—was still digging out of the war. Small cars were selling like gangbusters, but big money was scarce. Max Hoffman, BMW's importer in New York, had asked for the 507, but even relatively wealthy U.S. buyers balked at the price.

Also, there was competition. Mercedes-Benz had its lovely 300SL Gullwings and roadsters—cars with a tough, race-proven image—up against the cheerful, crisp lines of the 507, which suggested the good life on the boulevard more than the seriousness of the racetrack. And then you had Ferraris, Maseratis, Aston Martins, Porsches, Healeys, Corvettes, etc. The era was not hurting for interesting choices, if you had $8000 to throw around. So you would have to say, in the bluntest terms, the 507 was a sales flop for BMW.

But none of that takes anything away from the beauty—or technical prowess—of the 507. On the contrary, it just adds to the car's continuing exclusivity (my Sports Car Market price guide gives the 507 a current value of $280K–$385K) and explains its high standard of finish.

The body was designed by a German count, Albrecht Graf Goertz, who fled Germany before World War II, served in the U.S. Army in the Pacific theater, and then became an apprentice to Raymond Loewy after the war. Goertz started his own design firm in New York, caught the attention of Max Hoffman and was recommended to BMW as a designer for the new sports car.

And beneath that beautiful body was a chassis and running gear of

considerable sophistication. The engine, a 3.2-liter (3168-cc) all-aluminum ohv V-8 with two Zenith 2-barrel downdraft carburetors, was smooth and light, putting out 150 bhp at 5000 rpm, and mated to an all-synchro 4-speed manual gearbox. The double front A-arms and live rear axle (located by a Panhard rod) were sprung by torsion bars parallel to the bottom of the boxed frame rails, which had tubular crossmembers. The aluminum body featured the traditional BMW twin-kidney grille, but the kidneys were turned untraditionally sideways into a sleek low nose. BMW got an aerodynamically modified version (hardtop, undertray) to go 132 mph on the *Autobahn* from Munich to Nuremberg, while stock versions normally achieved around 120 mph.

Out of respect for our 50-year-old cars, our BMW hosts asked us not to drive quite that fast on the two-day drive over the mountains to Como. "Remember," said Andreas Klugescheid from BMW's Mobile Tradition press department, "these are not modern brakes, so don't follow anyone too closely—especially not another 507," he laughed.

But it was no joke; BMW had a lot of money and sentiment tied up in these five cars. The weight of history was upon us, albeit with a sunny day in Italy and the top down.

For all these warnings, our red 507 turned out to be easy to drive, without a lot of special allowances for age. Not only does it still look contemporary, but it's a remarkably modern-feeling car. The engine starts with a turn of the key and a couple of pumps on the accelerator pedal and revs in smooth, growly bursts that sound almost like a big electric drill being gunned. At idle, it has none of the lope of an American V-8 of the period. Handbrake release is an umbrella handle up under the dash, and the clutch is moderately light. Snick the precise-feeling gearbox into 1st and you're off.

Ride is civilized and compliant, with a bit of body roll in the corners, but good grip and light steering. The 4-wheel drum brakes aren't neck-snapping, but the pedal is firm and the car stops predictably and reasonably well from speed.

And speed isn't hard to generate. The 507's V-8 is remarkably flexible, pulling like a truck from 2000 rpm on up, but revving out smoothly and sweetly to its 6000-rpm redline. It's a beautiful engine, tailor-made for Alpine roads (or Italian foothills), because you don't have to stir the gearbox to put the power down through corners. You can leave it in 3rd for miles of moderately tight curves, and you seldom need to rev the engine past 4000 rpm.

And while you're rolling, the 507 cockpit is a nice place to be. The black leather seats are comfortable, if not heavily side-bolstered, and you look out over a four-spoke white plastic steering wheel and an elegant painted dash with white plastic knobs for heater, lights and the Becker-Mexico radio. A right-hand turn signal stalk takes some getting used to, and our only other ergonomic complaint was a worn 2nd-gear synchro that required accurate (or attempted accurate, in my case) double-clutching.

Our white plastic heater knobs were useful, too, as late afternoon brought swirling dark clouds and cooler weather—"the uncertain glory of an April Day," as Shakespeare's Proteus would say. It was collars up and full blast heat until we got to our hilltop hotel, l'Albereta in Erbusco, near the foot of Lago d'Iseo.

Rain and lightning borrowed from a Dracula film lashed our mountaintop redoubt all night, and the next day was cool and misty. Top-up time for the 507—a snug and easily erected top, with a pair of latches at the upper windshield. Ken and I also tried the 503 coupe, which proved a slightly softer and more civilized version of the 507, and a little less powerful with its lower compression (140-bhp) engine.

This car had a 4-speed column shifter with an almost delicate feel that allowed you to shift through the gears with a very light touch—much less clunky than the 3-speed column shifters of my youth. Or, at that age, maybe it was just me.

After lunch, we traded cars for the black 507. As we left the parking lot, Ken said, "Wow, this thing has much looser steering than the red car did. It really wanders."

"Feels like the rear axle is steering the car," I said, queasily. At that moment there was a metallic clatter under the car. "Sounds like something fell off," I remarked.

Ken immediately pulled over in a little village, and the rear end shifted with a loud clunk. We looked under the car and saw that a large nut had fallen off a locating arm on the rear axle. BMW's chase truck couldn't find it, so the 507 was abandoned at a village garage and we hitched a ride into Como with a journalist from Germany in a 5 Series. BMW flew a nut down from Germany and fixed the car the next day.

By that time, we were ensconced in our hotel in Como and riding a boat-taxi across the lake to the concours at Villa d'Este. This is a grand old hotel on the lakeshore, originally built as a home and pleasure palace for Cardinal Tolomeo Gallio in 1568, and since home to a succession of princes, dukes and other bewigged types who were not exactly on food stamps. It became a hotel—one of Europe's most famous and lavish—in 1873. Single rooms start at about $600/night; sort of like a Motel 6 with the decimal misplaced. But the grounds are beautiful, and so was the concours.

The *Concorso d'Eleganza* at Villa d'Este is a prestigious event in Europe. It was started in 1929, and is the oldest surviving concours in the world. Now sponsored by BMW, it's built on the concept of showing a small number of cars of great importance, as well as demonstrations of new technology and concept cars. So, essentially, you had 52 vintage entrants and a variety of show cars that ranged from minivans apparently designed for Buck Rogers' growing family and flying saucers with wheels to svelte aerodynamic prototypes. Also featured this year were the cars of Pininfarina (75 years' worth) and of course the BMW 503s and 507s. BMW also had a large display of Isettas—also celebrating their 50th.

We all have favorite cars at these affairs to which we keep returning, and mine were a stunning 1951 Ferrari 212 Vignale Spider, owned by American Jeff Fisher; a 1954 Zagato-bodied Fiat 8V and a 1964 Lancia Flavia Sport (also with Zagato body) belonging to Claude Picasso. I also hung around David Sydorick's 1961 Aston Martin DB4 GT Zagato long enough to make a pest of myself. Somebody must have slipped Zagato tablets in my champagne the night before.

Apparently, others shared my fascination with the Ferrari 212, as it won the Corrado Millanta Trophy, chosen by popular votes of the journalists and photographers present.

Sunday dawned with hardly any dawn at all, gray, cold and rainy. A shame, because the enlarged car show on the surrounding grounds was filled with lovely cars parked under plastic tarps. The crowds gathered in a large pavilion, shook off their umbrellas, drank espresso and watched a huge TV screen as Alonso held off Schumacher to win the San Marino GP. BMW also held a symposium on the enduring influence of 1950s' design in our lives, a fitting subject with 507s, 503s and Isettas parked everywhere around us.

Posed in the pavilion lobby as well were a pair of one-off alternate designs for the 507, a second-series 507 study done by Giovanni Michelotti, and the original 507 prototype done by Ernst Loof. Max Hoffman had reportedly examined the Loof car and said, "That will never sell." He then recommended Graf Goertz for the task of redesigning the body.

The Michelotti car contains many elements that would later turn up in the TR-4, and the Loof design is pleasing in its own way, but has a heavier and more Teutonic look—a little more like the Mercedes 300SL, but with a less becoming grille and nose treatment. I think BMW chose well, and the 507 still looks best as imagined by Goertz. Who, sadly, was in frail health and not well enough to attend this event.

Would that history, economics and the alignment of the planets had allowed his design to sell a few more copies. Mass production would have been nice. Then, a lot of us with less than $385,000 to spend might have one in the garage, even now. It's a car that still drives well and looks beautiful after 50 years, even if building it to such a high standard almost broke the company.

But it didn't, and all's well that ends well.